CO-OPERATIVE ENVIRONMENTAL GOVERNANCE

ENVIRONMENT & POLICY

VOLUME 12

The titles published in this series are listed at the end of this volume.

Co-operative Environmental Governance

Public-Private Agreements as a Policy Strategy

Edited by

Pieter Glasbergen

Utrecht University,
The Netherlands

KLUWER ACADEMIC PUBLISHERS
DORDRECHT / BOSTON / LONDON

A C.I.P. Catalogue record for this book is available from the Library of Congress.

ISBN 0-7923-5148-7 (HB)

Published by Kluwer Academic Publishers,
P.O. Box 17, 3300 AA Dordrecht, The Netherlands.

Sold and distributed in North, Central and South America
by Kluwer Academic Publishers,
101 Philip Drive, Norwell, MA 02061, U.S.A.

In all other countries, sold and distributed
by Kluwer Academic Publishers,
P.O. Box 322, 3300 AH Dordrecht, The Netherlands.

Printed on acid-free paper

TABLE OF CONTENTS

PREFACE

New philosophies of environmental management are being put to the test in many countries. New ideas are needed to replace or at least flank the old command and control approach, which has lost its credibility. One of the most interesting new avenues is co-operative environmental management, whereby public and private parties work together to tackle a problem. It is interesting because it seems to be well suited to handling complex environmental problems.

This kind of management makes use of the policy instrument known as the Environmental Agreement. That tool is geared to the development of sustainable *procedures* for working out solutions. The Environmental Agreement provides scope to deal with some essential characteristics of current environmental problems. Indeed, one of the most vexing aspects of environmental problems is uncertainty, both in the ecological sphere and with respect to the economic effects of intervention. In short, this instrument takes the unknown into account.

This approach has produced some good results and seems to have potential for more, though many questions remain. Answers can only be found by conducting thorough research. All the questions that have been raised refer in some way or another to the nature of environmental governance by agreement, the preconditions for this approach, and its chance of success. This book is an attempt to make progress by making sure that each step we take is a step in the right direction. It looks for answers in various places and from various angles. It takes an interdisciplinary approach to the task, analyzing theoretical issues and practical experiences in a number of countries.

The contributing authors set up their own basis for co-operation in 1997 during a workshop held at Utrecht University, entitled Negotiating Solutions to Environmental Problems. That event was made possible by a grant from the TERM Program, which resorts under the European Science Foundation. Since then, the participants have kept in touch, working out some of their papers to become chapters in this anthology. These contacts have also given momentum to my ongoing research program, Governing Sustainable Development.

On behalf of all the contributors, I wish to express my gratitude to those in charge of the TERM Program. In addition, I wish to thank Thea Schouten van

Regteren for her secretarial assistance all the way from the initial idea of organizing a workshop through the presentation of its outcome to a wider audience in the form of this book. I would also like to thank Nancy Smyth van Weesep for the English language editing. Finally, the authors and the editor would be grateful for any constructive criticism and suggestions from the readers.

Prof. Dr. Pieter Glasbergen
Utrecht University / Open University, The Netherlands

1. THE QUESTION OF ENVIRONMENTAL GOVERNANCE

PIETER GLASBERGEN

1.1 Introduction

Environmental issues have given new impetus to the debate on the opportunities and limitations of managing social change. The concept of governance presupposes a manageable society, or one that can at least be influenced. That concept is thus allied to a belief in progress; the malleability of society could be turned to its advantage. These elements have been pivotal to modern environmental policy from its very inception. Environmental change implies social change. The intent of environmental policy is to help shape a different society. No matter along which lines it is moulded, that society should be different and better to live in than the present one. Environmental policy is an attempt to induce people to change their behaviour and to imbue society with new and more ecologically sound social arrangements. These changes will have to be organized in one way or another. This raises questions as, who will manage that process and how? What are the options to get a process of environmental change going? Which criteria would the agents of change have to meet? To what extent can they live up to these requirements? The answers to these questions can be classified in terms of models of governance. Models of governance are developed to make complex issues more amenable to dedicated social action. They differ according to the role ascribed to government. However, that role is often a composite of standpoints on what makes environmental problems hard to manage. On the basis of different definitions of the governance issue, various ways can be advocated to introduce change.

In the quest for new roles for government, a new option has emerged: interactive forms of governance. It is based on what Jänicke (1992) has called the 'consensual capacity' as an institutional condition for success in environmental policy. The common denominator of these forms is co-operation between public and private parties. The parties commit themselves, through a more or less binding agreement, to work together to resolve specific environmental difficulties. Of course, co-operation is nothing new in itself. The novelty lies in its planned form. When co-operation is embedded in environmental policy, it becomes a

1

P. Glasbergen (ed.), Co-operative Environmental Governance, 1–18.

means to achieve the environmental objectives of the state. This new option goes by various names. Gray (1989) calls it 'collaborative management'. Van Vliet (1993) speaks of 'communicative governance'. And Glasbergen (1995) uses the term 'network management'. Recently, Lafferty and Meadowcroft (1996, p. 257) introduced the term 'co-operative management regimes' to embrace several of these concepts.

Applications of co-operative environmental management regimes are mainly found in West and Middle European countries. In any case, these countries have enough experience to carry on an empirically founded discussion on this kind of governance; the nature, the criteria that must be met, and the chance of making a real contribution to the task of dealing with environmental issues effectively.

The essays which make up this volume explore this new option in environmental governance. They all address the following key question: 'How to value public-private co-operation, especially environmental agreements, as a new policy strategy in environmental policy?' This introductory chapter leads the way by placing co-operative environmental management in the context of prevailing models of governance. The chapter closes with a preview of the structure of the book.

1.2 Environmental governance

The concept of governance refers to opportunities for goal-oriented and deliberate intervention in society. These opportunities purposefully determine the capacity to act (Kooiman, 1993, p. 46; Mayntz, 1993, p. 12). At an abstract and theoretical level, the concept of governance always encompasses three elements:

i)   The idea of an object and a subject in a specific relation.

      These relations are generally described in terms of dichotomies; for instance, government is linked to society, public to private, state to market. The ideas on governance are based on a certain conceptualization of these dichotomies. They identify the relevant categories of actors in civil society and define their roles.

ii)   The notion of an intent to induce change.

      This intent links the subject to the object. The way in which the intention is formulated usually expresses the concerns of the subject. These concerns may be shared by the object to greater or lesser extent. All measures that may be expected to lead to the intended result are formulated in accordance with that statement of intent.

iii) A conception of the social context in which the intent has to be realized.

      This refers to a more general set of ideas about what is needed for effective governance. Usually, these ideas will be a combination of specific object-bound notions and more general social principles. The latter might include political and cultural assessments and evaluations of the potential for change.

All ideas and concepts of governance somehow take a position on these three elements. To some degree, those positions are based on empirical observation. That is, they incorporate assessments of dependency relations, power positions, and opportunities to exert influence. These assessments form the basis for determining what *could* be. But normative considerations play an important role, too. Viewed in the light of concepts like democracy and the constitutional state, societal actors are expected to play specific roles and to fulfill specific functions. On this basis, it is decided what *should* be. Concepts of governance bring the empirical observations and normative considerations together in models. Models of governance are internally consistent sets of presumptions regarding how to bring about changes and why such methods are necessarily appropriate. They interpret social reality and imbue it with meaning in order to make goal-oriented action possible.

The above discussion refers to governance in a general sense. Environmental governance forms a specific category of that general notion. The field of environmental policy is rich in models of governance. We can mention five of them:
1) Regulatory control. This model assigns a key role to governments. They are seen as regulators of the process of change. In this model, setting standards and imposing stringent rules constitute the mechanisms of change.
2) Market regulation. This model assigns a key role to the price mechanism. It is seen as the means to achieve change. Governments are seen as facilitators of market processes. Their task is to ensure that markets can actually perform that function.
3) Civil society. This model assigns a key role to the self-confident citizens and the social ties they spontaneously create. The mechanism of change is perceived as an effort to make civil society more dynamic.
4) Contextual control and self-regulation. This model assigns a key role to the capacity for self-reflection in sub-systems of society. These systems have their own developmental logic. That internal dynamic can be put to use for collective problem-solving if governments create the right conditions.
5) Co-operative management. This model assigns a key role to collaborative relations between governments, mediating non-governmental organizations, and private interests. The mechanism for change lies in communication and dialogue, the results of which are laid down in voluntary agreements among the participants.

The models of governance are elaborated at different levels of theoretical explanation. Their origins and applications also vary. In practice, environmental governance will often consist of a mix of these ideas. Any application will have its own peculiar accents. The theoretical debate on governance is more concerned with ideas that may best be denoted as ideal types. Thus, they tend to be mutually exclusive. Any given model is usually a reaction to the limitations detected in another one. Each model has its own specific components and its own internal logic. At that level, these models can be reviewed, compared, and evaluated in terms of their potential contribution to environmental change.

1.3 Regulatory control

The basis for any environmental policy is regulatory control. Since the early 1970s, every developed society has had a continually expanding system of environmental legislation. New environmental goals are translated into legal rules. Each rule specifies conduct in which people should or should not engage. The rules are formulated in terms of prescriptions (what one *must* do) or proscriptions (what one *must not* do). If people do not follow the rules, they face negative sanctions.

The fact that the state uses regulatory control during the first stage of development of modern environmental policy is not surprising in itself. New social problems will always call for some form of regulatory intervention on the part of government. At the very least, a system of environmental legislation is needed as an infrastructure on which to build environmental policy. We can say that a model has coalesced when specific ideas on environmental governability crystallize on that surface. And this is indeed the case.

First of all, the idea has emerged that environmental problems emanate from acts that will only change when external force is applied. In line with this idea, environmental problems are perceived as the consequence of self-serving behavior on the part of actors in civil society. It is assumed that particular interests are so important, at least in the short term, and so far removed from the environmental interests that the actors should not be expected to come up with solutions to those problems. The state, as the institution that stands above the parties who are pursuing their own interests, will have to take the task of leadership upon itself.

Secondly, the idea of the manageable society has taken root. The idea is that society can be purposefully changed by government action, from a central position. The 'makability' stretches across the administration itself, covering both central and local/regional government, and reaches public authorities and private parties alike. Thus, the object of control by central government is not only the physical environment -the quality of which is in question- but also the social environment.

Under the regulatory control model, opportunities to reinforce environmental governability are sought in environmental standards. These are developed by the state and policed by its agencies. There is no doubt that this model has been reasonably effective. Indeed, many countries have achieved reductions in emission levels. Thereby, at any rate, they have turned the rising tide of pollution that has posed a direct threat to human health. At the same time, however, this model has been criticized. In fact, all of the models reviewed below may be considered as reactions to alleged shortcomings of regulatory control.

One of the key issues is whether or not this model has the capacity to resolve contradictions. Can it build a bridge between the desire for environmental quality, on one side of the credibility gap, and growth of business as a means of generating employment and national wealth, on the other side? For quite some time, it has been common practice to place business interests in opposition to environmental interests. Only lately has evidence come forth that strict regulation may also generate innovation. On the grounds of empirical cases, Porter and Van der Linde (1995 a, b), for example, demonstrate that properly designed environmental standards can trigger innovations that lower the cost of a product or improve its value. Stringent rules stimulate enhanced resource productivity - that is, more productive utilization of raw materials and energy, for example- and thus make companies more competitive, not less so. This standpoint has not escaped criticism altogether, however (see Palmer, Oates, and Portney, 1995). Nevertheless, it is clear that very few support the hypothesis that environmental regulations have by definition a large adverse effect on the competitiveness of industry.

Another key issue is whether or not regulation and hierarchy also work at a more developed stage of environmental policy. Those who doubt the positive effect of these methods are mainly influenced by a particular characteristic of the regulatory control model. Hajer has called that characteristic 'problem closure', which he describes as "defining a set of socially acceptable solutions for well defined problems" (Hajer, 1995, p. 21). The model is intrinsically reactive. This means that a system of rules can only be set up when a problem manifests itself in such a way that there is sufficient social support for developing policy to tackle it. If the problem is extremely complex, the expertise inadequate, and the level of uncertainty thus high, it is doubtful that central governance by way of stringent legislation will offer any perspective. In that event, regulatory control would appear to have reached a point of diminishing returns. The more elaborate the system of rules becomes and the greater the complexity of the problems that have to be managed, the less the effect will be. This phenomenon occurs in all (industrialized) countries that have a sophisticated environmental policy. In fact, it is one of the main reasons to develop other models of governance.

1.4 Market regulation

The neoclassical view is one of the reactions to the regulatory control model. The regulatory control model and the neoclassical view are based on a similar definition of the problem. Both see the central problem as a society that is not necessarily capable of spontaneously resolving its environmental problems. According to both models there is a divergence between public and private interests. The solutions offered by the regulatory control model -legal and administrative rules- are rejected, however. They are considered to be inefficient and not cost-effective. They are said to require a great -and often not very successful- effort on the part of governments and to result in solutions that (unnecessarily) restrict economic growth.

The neoclassical approach interprets environmental problems in terms of market failure (Baumol and Oates, 1988; Pearce and Turner, 1990). In very broad strokes, its reasoning may be sketched as follows. In a capitalist economic system, a price mechanism is used to regulate the use of products and services. In theory, the price of a good reflects the actual cost of producing it. This does not, however, apply to the environmental costs associated with production and consumption. The costs of more environmental degradation are not reflected in higher prices. Environmental degradation must be considered as an external effect, passed on from those who generate the effects to society as a whole. (For a discussion of the notion of external effects, see Löfgren, 1995; Van der Straaten and Gordon, 1995, pp. 143-148). For instance, those responsible for the external effects of acidification (such as motorists, farmers, the oil-refining industry) are not compelled to pay for the resulting damage caused to nature. In this model, market failure is blamed on the price mechanism. The so-called invisible hand fails to bring about the social optimum required to solve environmental problems. Once the financial incentive structure is changed, and environmental goods are rightly priced on the market -that is, to reflect their scarcity- public and private interests can be reconciled. In management terms, the task is to internalize the external effects.

Here we see that self-interest, defined as a materialistic interest, is not the problem -as it is in the regulatory control model- but the solution. If the intrinsic motivation for behavior is a materialistic one, that would be the direction in which to look for a way to bring about change. When the bill for negative environmental effects is presented to those who caused those effects, they will be motivated to seek out innovative ways of meeting environmental targets. In the words of Glasbergen and Cörvers (1995, p. 10), "Prices must tell the social truth about the cost of environmental degradation". This is what a financial incentive structure can achieve. When there is a change in relative prices (scarce resources and polluting activities are priced higher), people will do what they have always done. They adapt their behavior accordingly. They become more inventive and look for environmentally friendly alternatives. When people's self-interest is mobilized to solve environmental problems, government will have a much easier task; the market will take over the task of government.

The most significant instruments in this connection are those that conform to the market, such as taxes, fees, and tradeable permits. They signal to emitters the costs imposed on recipients. And they create incentives for emitters to take account of those costs in their use of natural resources (Moran, 1995, pp. 76-77).

Neoclassical notions have led to further theory-building and have spawned various schools of thought along the way. Much of the outcome, however, is abstract theory. In comparison, the number of applications in modern environmental policy is (still) low. To date, there is insufficient empirical basis for the supposed superiority of this governance model (see also Jänicke and Weidner, 1995, Chapter 1). Meanwhile, the approach has been showered with criticism. The critique is directed at the problem of setting prices when the degree of scarcity is unclear, at the difficulty of optimization, and at the inconvenience of using aggregation methods to determine preferences (Dietz, 1994, chapter 3). There is also the problem of valuing present resources in terms of future generations. Besides these technical drawbacks, it should be kept in mind that even though financial incentives are often presented as alternatives to regulation, they are really just another form of it (Jacobs, 1995, p. 115). Like the first model, this one also takes into account the power of the state, though less directly, by introducing taxes or tradeable permits. To be successful in reaching environmental objectives, financial incentives will always entail a strictly enforceable regulatory component, based on predetermined objectives. To diminish the total pollution permitted over time, governments should be active in keeping track of the achievements. Occasionally, they need to tighten the financial incentives.

1.5 Civil society

The regulatory control model puts its faith in the role of the state. The economic control model is more confident in the operation of market processes. Our third model is mainly based on trust in reasonable, self-confident citizens and their (spontaneous) social ties. In particular, the model is grounded in ties that are not built around a specific socio-economic interest. This model stresses discussion and debate as well as opposition and social critique as mechanisms for change.

The basic assumption is that a natural congruence may be expected between democratization and sound substantive environmental policy (Lafferty and Meadowcroft, 1996, p. 2). Inspiration for this model is drawn from various observations.

The first one is that from its inception, environmental policy has developed in tandem with social protest movements. By forming an oppositional force, the environmental movement has been able to unmask the failure of the state and/or the economy. There is no reason to believe that this will be any different in coming stages of environmental policy.

Secondly, a relation may be drawn between democratic values and the characteristics of present-day environmental aims. According to Paehlke (1996, p. 19), environmental policy objectives are essentially subjective and value-laden. This observation ties in with our earlier remarks about how to deal with the complexity of environmental problems. They cannot be tackled from a technological and rational approach, which entails setting off means against ends. The way ahead is to work toward an intersubjective consensus on what needs to be done.

Both observations suggest that wide public participation is imperative. Democracy is the process through which the environmental interest, as an interest common to the whole community, can be generated. In one way or another, this conclusion turns up in every discussion of the relevance of civil society to dealing with environmental issues. But in every instance, the emphasis is different. Two main streams may be distinguished: the participatory approach and the rejectionist approach.

Participation theorists emphasize the positive effects of what Fiorino (1995) has called the 'participation ideal'. The assumption is that participation at the very least contributes to the legitimacy of environmental policy. Another standpoint goes a step further, posing that participation may stimulate people to make links between their own behaviour and the common good (Witherspoon, 1996, p. 66). The most far-reaching standpoint is that only a participatory approach to policy-making can incorporate the needs of all segments of society, future generations, and other species (Paehlke, 1996, p. 19).

The rejectionist approach is more radical, in the sense that wide participation is perceived as an opportunity to correct the bias of the state and its agencies. Dryzek (1996, p. 122) formulates the approach thus: "the best strategy for ecological modernization begins in oppositional spheres in society."

The difference between these two streams lies mainly in the role ascribed to the state. The participatory approach sees the state as a positive force. The state is expected to create the conditions for an active public involvement in defining and tackling environmental issues. The state is seen as the only actor that can mobilize an unrestricted public debate and stimulate the formation of public opinion. Rejectionists, in contrast, take the opposite standpoint and criticize the role of the state. They perceive the state as an organ that supports the dominant private interests that block ecological change. In this approach, mobilization is geared toward the cultivation of a civil opposition force. And that force is to be directed against the state.

A point of criticism that has often been voiced is that this model is based on a somewhat naive complacency. It is assumed that the outcome of the public debate will be positive. But the outcome can just as easily be disagreement or actually run counter to the environmental values. It is thus questionable that the effort to mobilize civil society will actually lead to positive environmental results. On the other hand, the mobilization of civil society may lead to a more precise articulation of different conceptions of the public interest. In itself, this already enhances the quality of political decision-making by giving the discussion more breadth and depth. Apart from the model's own governing capacity, it also supports other forms of governance. And it is often applied for precisely that reason. For instance, it is used in the Local Agenda 21 projects, which are intended to reinforce public participation. Or, as in the Netherlands, it is used to give the organized environmental movement an institutionalized place as a partner in discussions on the development of environmental policy.

1.6 Contextual control and self-regulation

The fourth model is based on a fundamental critique of a single aspect of the regulatory control model. That aspect is the notion that society can be shaped by government intervention. The main source of inspiration for this critique lies in autopoietic social system theory and in the development of theory about reflexive law. According to the notion of governability couched in these theories, all hope is placed on self-governance within certain constraints (Luhmann, 1987; Teubner, 1992, 1993; Willke, 1991).

According to this model, the problems of governance derive from the specific characteristics of the objects of governance. In highly developed societies, where functional differentiation is extensive, there is no longer any reason to see these objects as persons or specific organizations. Rather, they should be sought among the sub-systems of civil society. Such systems have a particular character. On the one hand, they are mutually interdependent. But at the same time, they are extremely closed and self-organizing. Each system has an internal dynamic, which is geared toward maintaining and reproducing its own identity. These traits may be summed by the concept of self-referentialism. This concept refers to the tendency of organizations to observe their surroundings (and themselves) from their own perspective (Van Woerkum, 1997, pp. 13-14). They entertain their own idea of relevance and an entirely self-evident notion of what is important. As a consequence, a system will mainly interact with its surroundings when that interaction affirms the identity of the system and thus provides opportunities for self-reproduction. When such systems are confronted with information from the outside -for instance, laws that prescribe certain behavior- that intervention does not always have the intended effect. New inputs are transformed into something that has meaning in terms of the logic of the system itself. If the intervention does not fit the categories by which the system operates, it can't have the expected impact.

Here we see a variation on the theme of self-interest that played such a prominent role in the first two models with respect to determining a strategy for governance. But in this case, we are no longer concerned with the self-interest of individual actors. Rather, the self-interest at issue here is that of powerful systems within civil society. Precisely because those systems are powerful and robust, while being relatively closed, government policy is only able to penetrate them to a limited extent. Systems of this kind can only change through direct feedback from the environment. Commercial organizations do this through the market (Van Woerkum, 1997). The government, however, is unable to do likewise with its environmental aims. This type of system will necessarily have to be managed in a different way. The governance will have to be focused on stimulating the capacity for self-reflection and self-regulation.

This model presumes that the sub-systems of civil society -under certain conditions- are capable of taking responsibility for social problems. The systems have a the capacity to reflect. Therefore, they can compare their own qualities with those prevailing in the surroundings that they consider relevant. One of the necessary conditions for assuming responsibility for social problems is that the government must establish procedures and facilitate structures to promote dialogue between the parties with an interest in social issues. The term usually applied to that condition is 'concerted action'. This implies making arrangements that can lead to an effective integration of societal interests. Government intervention should also serve as a corrective measure in case the interests do not get equal consideration. In the background is the idea that equal treatment of the interests at play will make democratic learning processes possible. This determines the success of collective societal problem-solving through self-regulation (Huls and Stout, 1992; Eijlander, Gilhuis, and Peters, 1993).

In comparison with the previous models, this one is more abstract. Actually, it is a less stringent form of the regulatory control model, in the sense that it emphasizes contextual control. By doing so, the resistance to government policy can be mitigated and the potential of sub-systems to change can be stimulated. Only when such attempts prove to offer no perspective on change is it necessary to revert to more strict regulation.

Some applications go in the direction intended for this model. One example is the environmental care system developed for the chemical industry. Another is the system of self-regulation used by the nuclear power industry in the USA. Nevertheless, studies show that the scope for effective self-regulation is extremely limited. It can only operate under very narrowly defined circumstances. There must be substantial overlap between public interest and private interest; moreover, the number of players must be small and they must be united by a strongly perceived community of shared fate (Rees, 1994, p. 93; Gunningham, 1995).

One question that has been raised repeatedly is how to prevent self-regulation from degenerating into blatant protection of group interests. When there is too little centralized coordination, there is a risk that the most powerful parties will determine which rules will actually apply. At the other extreme, the system of legal preconditions may become so elaborate that the ideal of self-regulation will become an illusion.

1.7 Co-operative management

The fifth model also presumes that governments have only limited capacity to orient consciously social development. The central idea is that modern government can only make social change occur if the authorities are assured of co-operation by institutionalized private parties with a direct interest in the issues at stake. Their co-operation should be voluntary in nature and should take shape during negotiations between the parties most deeply involved.

According to this model, government policy should be taken as merely one of many factors that determine social developments. In market economies, governments have only limited opportunities to influence developments in the area of production, consumption, mobility, and technology. Many other forces are at play, and these are often stronger than government policy. However, it is precisely the (side) effects of these developments that determine environmental quality. Within the limited scope of government, environmental policy has a weak position. It is a policy that works reactively; that is, it kicks in when the damage comes to light. Furthermore, environmental policy often serves as a correction to other government policy (e.g., agricultural policy, or policy on traffic and transport) that supports environmentally unfriendly developments in society. In this context, public and private parties are highly dependent on one another in dealing with environmental problems. Policy has to be developed through complex interactions between parts of the state and parts of civil society. Sometimes those interactions are conflictual, sometimes consensual, but often there is a mix of conflict and consensus. In all situations, however, public and private parties need each other in order to find meaningful solutions.

The aim of the co-operative management model is to make the mutual dependencies productive. Governments should make use of the problem-solving capacity of private stakeholders by forging strategic alliances with them. Parties who are inclined to act independently must be stimulated to take concerted action in such a way that individual interests will be linked to collective ones.

A symbiotic relationship between governments and interest intermediaries has several advantages:
- It provides an orderly forum for interest articulation, while minimizing destructive conflict.
- Because interest groups provide data, interpretations, and perspectives, it is possible to enhance the range of policy alternatives.
- It leads to building up support for and legitimization of policy activities aimed at solving a problematic, as interest groups become partial stakeholders.
- To the extent that intermediary organizations are representative of their constituencies and are able to foster compliance on behalf of their membership, they have the potential to relieve government of significant regulatory burdens.

Communication and dialogue between parties that are considered equal are seen as the most important instruments. In order to rally a joint effort by public and private actors in dealing with environmental problems, structures should be developed for consultation and negotiation. These structures may be considered as temporary arrangements and in the course of negotiations they should promote collective action. The negotiations are supposed to arrive at a joint policy practice. Public and private parties bring in their knowledge and perceptions. Through negotiations, they try to formulate a common analytical and normative framework within which:
a) A shared definition of the problem emerges.
b) A common desire to tackle the problem evolves.
c) Each of the parties makes whatever contributions it can to the resolution of the problem.
Thus conceived, the negotiations entail a redefinition of group interests within the context of a broader common interest. The parties are partners in a collaborative attempt to resolve specific environmental difficulties. The results of the negotiations are laid down in the form of agreements and provide the basis for contracts that are more or less binding. In some cases, they are called voluntary

agreements, in other cases negotiated agreements. A more neutral term is the one used by the European Union, namely environmental agreements.[1] Co-operative environmental management is a rather young model of governance. But, as Jänicke and Weidner observed, there is a clear trend in various countries towards the participation of a broad spectrum of actors and voluntary co-operation of polluters. In their view successfull environmental policy depends more and more on this kind of management (Jänicke and Weidner, 1997, p. 308). Still unclear, however is how far environmental agreements may extend, what there limits are in terms of type of environmental problem, administrative regime, legal requisites etc.

1.8 A critical review of the models

Up to this point, the question of environmental governance does not seem to be very straightforward. At any rate, models show a wide range of theoretical points of departure. Each of the models discussed here posits a different imperative: 'because there is no other way…'. Furthermore, each of the models cites different reasons for success: 'because this is for those reasons the best approach to environmental problems …'. When we examine these models more closely, two issues in particular come to the fore.

First of all, we notice that the models are based on different disciplinary backgrounds:
- Lawyers tend to stress the importance of regulation.
- Economists tend to stress the importance of financial incentives.
- Political philosophers tend to stress the importance of civil society.
- Critical lawyers and political scientists tend to stress the model of contextual control and self-regulation.
- Co-operative approaches are favored in political science too.

All in all, not only do the models emanate from different disciplinary propositions; even within one and the same discipline, the question of environmental governance has led to different perspectives.

Secondly, we notice that different models emphasize different forms of rationality (see also Glasbergen, 1992, p. 198). The regulatory model is built on faith in functional rationality. On the basis of objective and true knowledge, a government would be able to put a design for a different society into practice.

---

[1] We prefer the last term. It remains to be seen how voluntary the commitments are. This is certainly questionable when agreements have been made under threat of interventionist legislation. Negotiation always takes place, even in the regulatory control model. Though important, the role of negotiation is not a distinguishing characteristic of the model.

This model is presumed to be effective among law-abiding citizens who respect authority. In the market model, faith is put in economic rationality. This form of rationality appeals to what has been called 'rational egoism' (Self, 1985, p. 54). Rational conduct is directed toward achieving one's own needs by the most efficient means. This model appeals to the 'homines economici', or the egocentric utility maximizers. The civil society model is grounded in communicative rationality. It revolves around reaching agreement on the basis of sound arguments and (mutual) understanding. In calling for reason and collective interests, it appeals to the altruistic social personality. Autopoietic social system thinking does not posit an all-embracing rationality. Instead, it emphasizes the partial rationalities of self-referential entities. The rationalities are derived from the specific functions that are performed. The co-operative management model is based on combinations of these types of rationality. In fact, it stresses the importance of creating cross-links. The state is assigned its own specific role. But there is also room for organized private actors in the process of policy development and implementation. Communication that leads to contractually agreed forms of self-government is an option. However, it is not entirely clear how the mix of rationalities would actually turn out.

These observations lead us to a more general conclusion, namely that the governance debate is highly normative in nature. Nonetheless, it is worth trying to clarify the options. That attempt leads to the following summary. Regulatory control is the logical starting point for any environmental policy. However, their are reasons, mentioned earlier, to assume that a further elaboration will reduce the opportunities for governance. The other models may be interpreted as attempts to capture the relation between public and private realms in another way. Assumptions about the 'steering' position of the state are a major reason to conceptualize that relation anew. In this respect, the market model might be the most vulnerable of all. It is an exquisite logical creation, but in practice it has only been tried in a limited way in certain contexts. The problem is that it is based, albeit implicitly, on the existence of a strong state that is able to govern by offering financial incentives. The civil society model can hardly be considered an independent model of governance, as was pointed out earlier. However, the strategy that is proposed in this model performs an important supporting function when governance is conducted according to the other models. The model that champions contextual control and self-regulation as the best strategy for resolving problems can be seen as a variation of the regulatory control model. The state continues to play a key role. The difference is that direct governance is replaced by governance from a distance. The co-operative management model clearly offers the most distinct alternative. This is partly because of the way it tries to combine various kinds of rationality. In principle, the model does not dispute the power of the state. Instead, it assigns the state different tasks, partly in relation to the different role played by private actors in tackling environmental issues.

Governance according to the co-operative management model is explicitly based on the uncertainty and complexity inherent in modern environmental problems. In seeking a resolution for these problems, this model takes a positive approach to complexity and uncertainty, defining them not as obstacles but as part of the challenge. The state and the articulate citizenry take up this challenge in an organized form of concerted action. In this sense, we consider this model to be the most viable alternative in the quest for environmental change.

1.9 The scope of this volume

The chapters in this book investigate the nature and preconditions of public-private co-operation and assess its chances for success. In particular, they examine the viability of environmental agreements as a policy strategy. The most remarkable shift that has taken place with the introduction of environmental agreements into environmental policy is that private parties have gained a more or less permanent place in the development and implementation of environmental policy. But the implications run deeper. This governance can also be seen as a fundamental shift in the ideology of liberal democracy (see also Van Gunsteren, 1994). In principle, this ideology does not recognize the role of private actors on the public stage. They may be consulted and they may be engaged in executing policy. But as political actors, they do not exist conceptually. Liberal democracies draw a line at some point between public and private realms. That boundary separates areas of accountability. Of course, the parties regularly overstep their bounds. Still, private actors are not expected to share in the political functions. Co-operative environmental management tries to make the problematic relations between liberal-democratic politics and managerial capitalistic economic activity easier to handle. Private parties are explicitly assigned a role in the public arena. They are active participants in public decision-making. And they take responsibility for implementing these decisions.

But their active involvement creates a few problems as well. For instance, there is some danger of regulatory capture. This may occur if information asymmetries are not eliminated in partnership relations. Partnership is then used to secure the particular interests of individual parties. Or it is seen as a very gentle way to make use of obstructive power. Another danger is that public issues may become commodified. This happens when representatives of the state come to see their role as only one party among many in a bargaining process in which they are involved as stakeholders. In both situations, the attention may shift away from the public goal to the specific problems. Thereby, the negotiations may become bogged down in the individual parties and their conceptualization of the problem. This is a manifestation of the tension between the effort to form contractual relations, on the one hand, and democracy -in the sense of widespread debate about issues of public concern- on the other.

In light of these critical remarks, the question arises whether public and private parties can really be equal. Governments perform specific tasks: they provide for rule of law, a guarantee of social justice, etc. The government may be just one of the many parties that govern, but it will always have a special status. Due to the nature and function of governments, they are not just another market party (Kickert, Klein and Koppenjan, 1997, p. 177).

Despite these remarks, it is clear that the governance model that we examine here has the potential, at least in principle, to bridge the gulf between the economy and the environment. Modern environmental policy is based on the standpoint that these two areas should be joined. Co-operative environmental management is a model that seeks to operationalize that goal in a modern form of governance.

1.10 The structure of the volume

The extent to which public and private interests can be reconciled in environmental governance -and under which conditions- is examined in this book. And this question is considered from various interdisciplinary perspectives. First of all, attention is focused on the experience that has been gained thus far in applying the model of co-operative management. How does it work, and when doesn't it? A second concern is how this model relates to other forms of governance. Are the models interchangeable? A third concern pertains to the social context of the co-operative management model. Does this model suit particular political cultures and structures better than other ones? This book will not be able to provide a definitive or exhaustive answer to these questions. But it will make a good start in that direction.

The volume opens with a series of chapters on the more general aspects of co-operative environmental management. The focus then narrows to a number of country profiles, each of which is written from the perspective of a particular discipline. Then, a number of more general evaluations draw the volume to a close.

In part one -*conceptualizations*- James Meadowcroft picks up where this introductory chapter leaves off. He reviews the potential advantages and considers some of the more obvious objections to co-operative regimes. Advantages are related to pluralist inputs to policy-making; transformations of interest; flexibility; stable and legitimate outcomes; and enhanced learning processes. The objections are considered along four lines of argument: power, democracy, efficiency, and political culture. Two aspects of the problematic are then worked out by Huib Ernste and Yrjö Haila respectively.

Huib Ernste draws attention to the parallels between the management styles in private industry and public administration. He observes a common ground. Ernste argues that it is not so much the introduction of market principles but rather both internal and external co-operation that constitute the innovative core of new management paradigms. He illustrates his argument with examples from Switzerland.

The focus of Yrjö Haila's chapter is on the potential of socio-ecological analysis for deliberative environmental policy. The notion of 'eco-social complex' is introduced and elaborated to serve as a conceptual tool in this endeavour. He concludes that democratic delibetation would be valuable in two ways: a) in opening up the process of problem definition, and b) in widening the circle of citizens involved.

In part two -*experiences*- Betty Gebers opens with a general overview of the design and use of environmental agreements in nine countries. She shows that, in spite of all diversities with regard to legal form and integration into the legal system, some similarities can be observed. This applies to the motives, the significance for environmental policy and the policy fields in which it is mainly applied.

Seyad, Baeke and De Clercq explore the economic-institutional context in which environmental agreements are applicated. Through a comparative case-study of Belgian environmental agreements they determine which factors influence performance in the sense of effectiveness, efficiency, and transparency. Such factors are a.o. traditions of joint-policy-making a small group of negotiators, a real economic advantage, the use of alternative policy instruments.

Pieter Glasbergen analyses the application of environmental agreements in the Netherlands from a learning perspective. At first, the instrument was mainly used as a marketing device in propagating environmental policy. Since then, it has turned into a full-fledged new concept in Dutch governance. The convenant now specifies the elements that are crucial to making progress in learning. On the basis of the experiences some prerequisites for co-operative environmental management are formulated.

Heidi Bergmann, Karl Ludwig Brockmann and Klaus Rennings review German practice from a neo-liberal angle. The analytical grid encompasses the criteria 'goal conformity', 'system conformity', 'economic efficiency' and 'institutional controllability'. They conclude that environmental agreements require standards and a framework set by government. Once such a framework is set, agreements may be used as a tool for implementation within an environmental policy mix.

Ellen Margrethe Basse analyses the legal aspects of the Danish experiences with environmental agreements in the area of waste. Her specific focus is on the relation between the use of contracts and trade-freedom. She lists some criteria that can be used to evaluate future-potential of environmental agreements from the perspective of trade-law; such as exclusivity, incapacity, and physical scale of the contracts.

Martin Enevoldsen relates the practice of environmental agreements in Austria, Denmark and the Netherlands to theory of democracy. In his chapter he follows a normative approach. He concludes that major improvements are necessary to make the practices live up with the requirements of democracy. He closes his chapter with possible avenues for improving the democratic quality.

In part three -prospects- Andrew Blowers explores the political and administrative boundaries of ecological modernization as a broader approach environmental agreements fit in. The argument put forward here is that EM offers, at least, only a partial solution to the problem of securing sustainable development. According to his view the kind of changes that will be necessary to deal with the environmental question are: a) commitment to long term environmental management; b) commitment to social equality; c) commitment to political participation.

In the concluding remarks Peter Driessen, with all the information of foregoing chapters in hand, discusses co-operative environmental management again. The aspects he looks at are: the game, the players, and the scope of the game. He closes with the expectations we can have, and certainly not have, of governance by agreement.

Part I

Conceptualizations

2. CO-OPERATIVE MANAGEMENT REGIMES: A WAY FORWARD?

JAMES MEADOWCROFT

2.1 Introduction

This chapter will explore the potential which a particular pattern of interactive decision making offers for the successful management of environmental problems in industrialised democratic states. The approach involves drawing together partners from different sectors of social life to collectively define and implement solutions to specific environmental challenges. Such negotiated environmental settlements can be described as 'co-operative management regimes' (Lafferty and Meadowcroft, 1996), but their character might also be captured by phrases such as 'collaborative environmental administration', 'environmental co-management', or even 'environmental corporatism'. Here it will be argued that such initiatives offer a promising alternative to traditional regulatory strategies in dealing with some of the more intractable environmental dilemmas.

The perspective explored here reflects the spirit of much recent writing on environmental management, emphasising negotiation and collaboration as essential to achieving effective solutions to environmental problems (Lee, 1989; Carley and Christie, 1992). It echoes the language of 'partnership' and 'stakeholding' associated with international forums such as the Rio Earth Summit, and with the national sustainable development implementation efforts undertaken in its wake (Grubb, 1993; Carew Reid et al, 1994; Voisey et al, 1996). It owes much to recent US studies on the negotiated resolution of complex social disputes (Susskind and Cruikshank, 1987; Gray, 1989) and to the path breaking work on 'network management' undertaken by a number of European scholars (Glasbergen and Driessen, 1993; Glasbergen, 1995).

More generally, the discussion should be understood within the context of recent interest in new patterns of governance in contemporary industrialised democracies - with the withdrawal of government from spheres of social life in which it had previously been active (privatisation, de-regulation), the devolution of responsibility for service provision from central ministries to semi-autonomous public agencies (governmental down-sizing, contracting out), and the reform of administrative structures and practices to embody private-sector values (internal markets, new public administration) (Marin and Mayntz, 1991; Kooiman, 1993).

P. Glasbergen (ed.), Co-operative Environmental Governance, 21–42.
© 1998 *Kluwer Academic Publishers. Printed in the Netherlands.*

Some writers have argued that the state should 'steer' and not 'row' (Osborne and Gaebler, 1992); others have referred to the 'hollowing out of the state' (Rhodes, 1996); and many have commented on the declining capacity of national government to orient societal development in the face of increasing complexity, financial constraints, sceptical publics and especially the 'globalisation' of economic power (through transnational corporations) and of political decision-making (via supra-national entities such as the European Union). In the words of one analyst, the past decade has witnessed increasing interest in approaches to "governing in terms of 'co'" such as co-steering, co-managing, co-producing and co-allocating" (Kooiman, 1993, p. 2).

This discussion of co-operative environmental management is organised in four parts. The first section considers the distinctive characteristics of cross sectoral collaborative approaches to managing environmental problems. This is followed by an analysis of the advantages that may be derived from the deployment of such mechanisms. Next comes an examination of some of the main objections lodged against this approach to environmental management. The final section of the chapter draws together the strands of the discussion, and suggests a number of issues for further investigation.

2.2 What are co-operative management regimes?

A co-operative management regime is a form of social regulation in which groups originating in different spheres of social life, and reflecting distinct perspectives and interests, participate in debate and negotiation to achieve a common understanding of a specific problem, and then implement a collective plan for its resolution. Central features of such mechanisms are that they:

- involve participants from more than one 'sector' of social life: they imply not just co-operation among business interests, or among pro-environmental groups, or among various government agencies and departments, but rather collaboration across these broad areas of social life.[1]
- rest upon the representation of organised interests rather than on individual citizen participation;

---

[1] While the term 'sector' is frequently used to denote a distinct domain of economic activity (for example, the oil and gas industry or agriculture) it is used here to refer to three major 'spheres of social life': a) the 'public' sector (government at all levels, official agencies and semi-autonomous regulatory bodies) b) the bussiness sector (firms, business organisations); and c) the 'not-for-profit'/NGO/'civil society' sector (environmental groups, community action groups). In the expression 'public/private co-operation', 'private' refers to actors drawn from either of the two non-governmental sectors.

- depend upon processes of discursive consensus formation - it is through a shared experience of attempting to come to terms with a complex issue, through exchange and interaction among participants from quite different backgrounds, that the working-group comes to construct a common understanding of the problems to be addressed, and of the nature of potential solutions.
- require each partner to join in carrying out an agreed solution: at the minimum each organisation must publicly endorse the outcome and take up the task of 'selling it' to its membership; but groups may be more directly involved in 'implementation' through the assumption of legal obligations, and/or the devolution of administrative authority from governmental agencies.
- include a framework for review of the original agreements in light of practical experience.
- involve a significant cross-section of the groups and interests implicated in a particular problem nexus.

The expression 'co-operative management regime' seems appropriate to this form of negotiated environmental problem solving because such arrangements imply an attempt by organisations based in different domains of social life to collaboratively manage a particular set of environment-related impacts over a period of time. From the point of view of government they represent an attempt to reach beyond traditional modes of governance to develop new forms of public/private linkage which may enhance the societal potential for mediating environmental conflicts, managing environmental risks, and resolving environmental problems. 'Co-operative' emphasises that the processes are group based, and depend upon a collaborative approach to problem definition and solution. 'Management' indicates that these mechanisms are practically oriented; they are not just 'talking shops', 'think-tanks' or 'consultative forums'. From the perspective of the participating organisations they represent a means of managing a specific range of environmental impacts; but from the point of view of society more generally they may be viewed as components of a wider meta-strategy of environmental management. 'Regime' suggests (perhaps over-grandly) both that such initiatives constitute patterns of interaction for regulating a domain of environmental behaviour over time, and that their legal status may be somewhat uncertain.

It is important to note that co-operative management regimes are *not* envisaged as a replacement for the entire structure of existing environmental regulation and government environmental policy making - although in specific contexts certain forms of policy development and regulatory control may be devolved to co-operative management bodies. On the contrary, co-operative management could only represent a coherent response to the challenge posed by environmental problems when operating within a clear framework established by central governmental initiatives. Nor is co-operative management incompatible with the deployment of market-based environmental incentive systems. Rather it is a flexible management mechanism which can be grafted onto existing systems of policy-making and administration. Such regimes are not intended radically to displace other forms of interest accommodation and political representation, but rather to supplement these - although it is true that over time the wide-scale development of co-operative management mechanisms could significantly alter patterns of social decision-making.

Interest group pluralism, corporatism and policy networks
One way to appreciate what is distinctive about co-operative management is to consider its relationship with three ideal-type descriptions of policy formulation processes which have been exhaustively discussed in the politics literature. Classic *interest group pluralism* understands political decisions as resulting from the interaction of organised societal interests within the political system (Dahl, 1971; Polsby, 1985). Groups and interests are many, varied and overlapping, so no one interest can gain a stable monopoly of political power. Structural circumstances may help or hinder the articulation of group perspectives, but given the appropriate stimulus even the most diffuse or disadvantaged interests can exercise some political influence. For the most part, the work of politicians and bureaucrats is to be understood in terms of balancing and reconciling conflicting interests. Traditional *corporatist* approaches emphasise bargaining among the most powerful organised interests to conclude 'peak agreements', which are subsequently endorsed by the formal political institutions (Schmitter and Lehmbruch, 1979; Cawson, 1986). This highly structured and relatively closed system of interest intermediation is based around the key economic actors - business and labour - and government itself. It depends upon broad-based but hierarchically organised groups whose leaders can commit their members to comprehensive accords. The more recently developed *policy network* model emphasises the existence of stable policy communities which determine the development of government activity in specific issue domains (Marsh and Rhodes, 1992; Smith, 1993). Networks may be relatively closed to outsiders, but there are many different networks (specialised in particular policy domains), and over time network configurations change. Different kinds of groups may be involved in varied networks, with scientific or technical experts, for example, playing a role alongside professional organisations and economic interests.

Like the pluralist model, co-operative management highlights the potential input to policy-making of a wide range of social actors. But it differs in its reliance on problem-specific institutional frameworks - rather than established political/electoral procedures - to construct consensus and effect change. Like the corporatist model, co-operative management depends upon direct negotiation among concerned interests and collaborative implementation of accords. But it relies on opening this process beyond a handful of 'peak' economic organisations, and on the constitution of different forums for different decision-contexts. Like the policy network approach, co-operative management rests on the creation of matrices specific to each issue area. Yet it implies a more open process - both in terms of wider group participation, and transparency to outside scrutiny. Moreover it requires a broadening of interaction from a contribution to the formulation of government policy, towards the conclusion of formal agreements for joint implementation. Unlike the standard versions of these established models, collaborative management emphasises discursive consensus formation and the reconstitution of interests, rather than simply negotiation and compromise among pre-established interest perspectives.[2]

Cross-organisational collaboration and negotiated dispute settlement
Another way to understand the notion of a 'co-operative management regime' is to consider how it relates to broader ideas about cross-organisational collaboration and 'negotiated dispute settlement'. The American organisation theorist Barbara Gray has developed an approach to collaborative interaction that emphasises 'emergent interorganisational processes', and understands the rise of negotiated orders as a response to turbulent exogenous conditions confronting organisational actors. Gray distinguishes i) collaboration initiated by a common perception of a problem from that prompted by conflict, and ii) collaboration designed to secure informational exchange (perhaps leading to further voluntary co-operation) from that intended to produce a binding joint implementation accord. The result is a four-fold classification of collaborative designs: a) 'appreciative planning' (shared vision/information exchange); b) 'collective strategies' (shared vision/joint agreements); c) 'dialogues' (conflict/information exchange); and d) 'negotiated settlements' (conflict/joint agreements), (Gray, 1989). The co-operative environmental management regimes considered here are obviously a more specific form of interaction than the comprehensive portrait offered by Gray. In the first place, they are constituted around the particular problem domain of environmental behaviour. Second, they necessarily involve cross-sectoral collaboration, that is to say participant groups will not all originate from the same sphere of social life. Third, since we are concerned with public/private interaction, at least one partner to the collaborative interaction will normally represent the public (governmental or state) sector. Finally, co-operative management regimes imply not just a 'one-off' agreement, but embody some

---

[2] As an ideal-type, co-operative management has some affinity both with the 'corporate pluralism' discussed by Rokkan (1966) and the 'associative' perspective on democracy developed by theorists such as Hirst (1994).

intention to extend the collaborative process forward over time. Co-operative management appears to fit most obviously into Gray's fourth category - 'negotiated settlements' (conflict leading the partners to search for a joint agreement). In fact, it straddles the second and fourth categories ('collective strategies' and 'negotiated settlements'), suggesting that the distinction between collaborations motivated by perception of a common challenge and by conflict are less easy to disentangle that Gray assumes. For in the arena of cross sectoral environmental collaboration, partners are almost always conscious *both* of conflict (or potential conflict) and of common concerns.

The literature on negotiated dispute resolution focuses upon social conflict, and explores advantages which may accrue to disputants (and to society more generally) from negotiated solutions to problems which would otherwise be resolved by judicial, administrative, or perhaps legislative closure. This literature is rich with insights into the circumstances in which parties are led to explore negotiated conflict resolution, and into the process factors which influence the conduct of such exercises. Standardly, distinctions are made among the various phases of the process (pre-negotiation, negotiation, implementation), and among different types of negotiative encounter according to the extent to which they depend upon the 'good offices' of outside interlocutors ('unassisted negotiation', versus negotiation involving facilitators, mediators or non-binding arbitration). Clearly all co-operative management regimes involve negotiation, although not every negotiation results in the structured, longer term, collaborative problem definition and implementation effort that can be referred to as 'co-operative management'. Discussions of 'negotiated settlements' (Susskind and Cruikshank, 1987) and 'environmental dispute resolution' (Rabe, 1988) traditionally assume a pre-existing conflict - where parties are organised or organising to seek redress through appeal to legally authoritative bodies - and negotiations are instituted to reduce costs and uncertainty, and to secure otherwise more favourable outcomes. In some contexts co-operative management can be considered a more pro-active strategy, one that may deployed *in advance* of the emergence of a clearly polarised conflict.

'Stakeholder' representation

It is also worth noting that there is considerable resonance between the concept of co-operative management and the fashionable notion of 'stakeholder' representation. Both apparently rest on the idea that groups beyond those that have traditionally taken decisions may have a 'stake' in an ongoing enterprise; that these parties are entitled to some 'voice' in matters of common concern; and that their participation may be necessary to generate successful outcomes (Freeman and Evan, 1990; Gamble and Kelly, 1996). Of course, the idea of stakeholding has been applied in various ways. Sometimes it refers to the governance of business corporations (where, in addition to the stock-holders, 'stakeholders' may be understood to include managers, workers, customers, suppliers, local communities, or even society as a whole). Sometimes it denotes more widely the range of constituencies effected by a particular public or private sector decision-context. And sometimes it applies to 'the stakeholder society' in which no group is excluded from the benefits of citizenship or from participating in society's decision-making. Since co-operative management takes specific 'environmental problem areas' as the focus for its activity it differs both from the first 'stakeholder' variant (which is centred around the firm and issues of corporate governance), and the third 'stakeholder' variant (which takes 'society' at large as the problem domain, and is concerned with politico/social exclusion). It is the second variant, where 'stakeholding' is conceptualised with respect to issue areas, and which resonates with the emphasis on 'major groups' in Agenda 21, which comes closest to the spirit of co-operative environmental management (UNCED, 1992). On the other hand, co-operative management envisages a more precise and closely defined form of collaborative interaction than is often denoted by the 'stakeholder' concept. It implies more than just consultation or negotiation among such groups, but rather a commitment to secure a collaborative intervention in practice.

Strategic policy tools
It is also helpful to consider co-operative management from the perspective of the range of instruments governments may employ to orient societal development along desired pathways. In terms of strategic policy tools, government officials are usually considered to confront a choice among mechanisms that rely on a) law, regulation and sanctions, b) taxes, prices and markets, and c) normative prescriptions and persuasion. Each approach has advantages and disadvantages which have been exhaustively discussed in the literature (Pearce and Warford, 1993; Eckersley, 1995). Legal prohibitions and injunctions can be brought to bear uniformly, with a high certainty of general compliance, but are ill-suited to address variety and complexity, may encourage allocative inefficiencies, and have high enforcement costs. Financial instruments which operate on the self-interested behaviour of economic agents typically have low enforcement costs and are assumed to avoid allocative distortions, but may be unable to circumvent structural constraints and may generate undesirable distributive impacts. Campaigns of education and moral persuasion are comparatively cheap and avoid a proliferation of regulatory controls, but are often of dubious efficacy. Within the environmental policy domain legal prohibition, imposition of maximum and minimum standards, permitting and inspection have formed the core of the system of governance built up in the developed countries over the past three decades. Economic instruments such as environmental taxation, tradable emission permits,

and deposit-refund schemes have begun to be deployed (Andersen, 1994). And normative injunctions - for consumers to 'shop green', or for producers to explore eco-efficiency options, for example - have occupied a regular, if subsidiary, place in the overall package.

Current interest in collaborative interaction and public/private partnerships suggest that another approach - d) organisational engagement, negotiation and co-operation - can been added to the standard list of environmental policy instruments. From the perspective of governmental officials and agencies, entering into direct dialogue with extra-governmental social partners and reaching agreement on a joint course of action can be understood as an alternative to established regulatory, financial, and educative modes of policy implementation. Yet while in one sense these options constitute alternative forms of intervention, they actually rest on complex interdependencies. 'Economic instruments', for example, operate through market mediated transactions, but they require law - changes to tax regulations and administrative rules - to make them operative. And markets have significant normative, as well as legal, underpinnings. For their part, co-operative solutions are always undertaken within a wider context of legal, financial and normative policy initiatives and frameworks; and such conditions, and changes to them, can make negotiated solutions more or less practical, and more or less effective.

Furthermore, while co-operative management arrangements can be conceptualised as an instrument for government policy delivery, they cannot be adequately conceptualised as *merely* a policy implementation mechanism. This is because co-operative management can be as much a way of 'making' policy (determining solutions) in particular environmental domains as a way of 'implementing' pre (and externally) fixed objectives. Central to such approaches is the idea that the parties come together to explore collectively the problem area and to arrive at a reasonable way forward. In other words, when government agencies engage in such interaction they must be willing to accept that their own perceptions of problems and possibilities will be supplemented, enhanced, and altered by the perspectives of other partners, just as they anticipate that the collaborative process will also shift the understandings and behaviour of the non-governmental participants. Perhaps it is most accurate to say that in order to achieve higher level policy objectives, governments initiate co-operative processes (involving collaborative problem definition, solution space determination, and practical action) which result in (second order) policy-making and implementation. Of course, these collaborative interactions may in turn feed back to influence the processes through which the higher order policy objectives and implementation strategies were determined.

Co-operative environmental management regimes are an innovative way of assuring public/private collaboration in resolving environmental problems. But not every private input to public sector decision making, nor every case of private sector involvement in public policy implementation can be considered part of co-operative management. Traditional lobbying and political bargaining, closed corporatist patterns of interest intermediation, and the operation of standard policy networks (whether rigid or permeable) allow private actors to influence policy outcomes - but they do not embody the discursive, collaborative, and inclusive norms associated with the form of environmental management discussed here. Traditional procurement and tendering procedures, as well as more innovative 'contracting out' of social service provision, may draw private actors into the administration and delivery of public programmes - but they do not necessarily involve the creative, cross-organisational interaction and problem-solving focus associated with co-operative management. 'Negotiation' and 'agreement' have long featured in all areas of public/private interaction, but the sorts of negotiations and agreements in which we are interested here are those that are part of a collaborative process in which partners from various areas of social life engage in a cross-sectoral, dialogic, iterative and practically oriented approach to managing environmental burdens.

2.3. Advantages of co-operative environmental management

There are many potential advantages to collaborative forms of environmental management. Let me suggest six major elements.

First, they provide *a structured framework for encouraging pluralist inputs* to environmental policy-making. This is particularly important in the environmental field because of the complexity of the interests that may be affected by shifts in environmental conditions and management practices, and the difficulty in predicting medium term impacts both in the bio-physical and politico-social realms. No one party or narrow grouping of parties possesses the knowledge necessary to resolve environmental disputes successfully. Thus a wide spectrum of representation encourages more effective policy-making. On the other hand, what is required are not just chaotic, but structured, inputs. If the policy-making process is simply thrown wide to 'all-comers', closure may prove impossible as any group may veto movement towards decision. Alternatively debate may become highly polarised, and the possibility of constructive compromise obscured. Yet here co-operative management comes into its own, because it generates a structured context within which a limited number of groups can collaborate in policy formation and implementation. In other words, it broadens the range of social actors involved in the environmental decision-making 'loop', but does so in a way that can focus attention on problem definition, encourage convergences of approach, and facilitate the search for achievable solutions.

Second, they provide *a mechanism for building consensus and more especially for transforming interests*. Environmental problems may call into question deeply entrenched social practices, and an effective solution to a particular dilemma may only be possible through a gradual alteration of established patterns of activity. Co-operative management can provide a context to facilitate such change. Reference has already been made to the notion of 'transforming' interests, but it is necessary to consider this more carefully here. In much contemporary discussion 'interests' are taken as essentially 'given' - as a fixed element of a problem structure (like consumer preferences in micro economic calculus). Moreover they are almost exclusively associated with material advantage. Thus political problem solving is reduced to reconciling disputes over the distribution of material resources among established interest coalitions. In fact, however, politics has always at least in part been about changing the ways in which problems are defined, and the most promising approaches to conflict resolution attempt to alter the parties' conceptualisations of their interests. As patterns of perceived interest shift, solution spaces can come to encompass previously unimagined outcomes. Co-operative management regimes encourage such processes: direct contact and repetitive interaction among the parties, mutual recognition of each group as a legitimate interlocutor, commitments to jointly implement agreements, and the extended time-frame, all contribute to this effect. As groups explore the dimensions of an issue together, their estimates of other actors (both inside and outside the working group), and their expectations for the future, will change. For example, a producers group with an initial brief to resist regulatory initiatives may become convinced of the importance of structural reform in its sector to assure long term environmental and financial viability.[3] Precisely because identities (and interests related to these identities) are plural, individuals and groups can to some extent redefine their characters. Furthermore, new forms of interest and identity can be generated through social interaction and empathic connection. It is true that deep structural factors underpin configurations of interest: a city built with shops and offices located far from residential areas shapes the interests of its commuters; and no spontaneous reconceptualisation of subjective identity is likely to change wide-scale dependence upon the automobile. Yet changes of problem perception can motivate groups and individuals to seek consciously the gradual adjustment of apparently immutable structural constraints. Again, the longer term approach of co-operative management encourages such initiatives.

Third, co-operative management regimes are highly *flexible*. They can be adapted to different circumstances and applied in different contexts. This can be

---

[3] Consider for example the experience of the Gelre Valley Commission in the Netherlands (Driessen and Glasbergen, 1995)

appreciated in various ways. First, with respect to the nature of the problem area around which the parties are assembled. This can take many different forms and may be based around: i) a region or locality (a depressed rural or urban area undergoing redevelopment); ii) a specific industrial sector (say the chemical industry or the cattle industry); iii) a production/consumption domain (the motor car); iv) a scarce resource (water management); v) an acute environmental effect (urban air pollution); vi) a realm of social activity (transport); vii) special events or programmes (holding the Olympics). This flexibility is advantageous because it means that such systems can be established in a piecemeal way as different problem areas are identified, and/or 'opportunistically' - when particular constellations of political and economic circumstance are favourable. Furthermore, in a context where we do not know which approaches to the problem of sustainable development are most likely to be successful (by region, by industry, by product group, by environmental media, etc.), this permits experimentation with different management strategies (Meadowcroft, 1997a).

Flexibility also exists in relation to the kinds of groups involved, the forms of debate and decision procedure, the management time frame, and the nature of the implementation stage. As we have already observed, groups are drawn from the governmental, business and 'not-for profit' sectors - but there is an immense array of organisations here. State representation may come from central ministries, autonomous regulatory bodies, or regional and local administrations; business participants may be individual firms, industry-specific lobby groups, or business federations; environmental and community organisations can range from nation-wide movements to local networks. Debate and decision procedures can be adapted to the particular problem and the needs of the participants. They may include different forms of encounter session, visits, commissioned research, and so on. They may involve contributions from a mediator or formal conciliation techniques. Time frames can also be adjusted to match the urgency of the problem and the scale on which it needs to be tackled. In general, however they will focus on medium time scales: at least four or five years initially, with a perspective towards a decade or two into the future.

Co-operative management regimes also can be considered flexible in so far as their establishment does not necessarily require substantial antecedent legislative and/or administrative reform. Formal institutional change can necessitate significant resources of time and money, and generate much resistance. Experience with major reforms at environment ministry or agency level suggests that consideration of substantive problems may often be delayed significantly during periods of organisational overhaul. Such problems can to some extent be side-stepped through co-operative management, because negotiating bodies can be established directly and some of the more formal dimensions of reform can be addressed at a later stage. This also means that co-operative management regimes can be dovetailed to a great variety of existing institutional set-ups.

Fourth, they have the potential to generate *more stable and legitimate policy outcomes*. Precisely because many relevant groups are involved in concluding and enforcing an agreed solution set, policies may appear more authoritative in the eyes of concerned publics, and there may be greater confidence in policy continuity. Enhanced legitimacy cannot just be assumed - for the range of groups involved, the way negotiations are conducted and the practical results of the implementation phase, will all impact upon public perceptions. The relative openness and transparency of the process is also clearly crucial to public acceptability.

Fifth, co-operative management arrangements can provide a context in which *expert scientific and technical advice, and specialist environmental assessment techniques,* can be introduced into the environmental decision process in a fruitful manner. Participants in the interactive negotiating process are in some sense 'elites' (mandated representatives of particular organisations) and they are better equipped to make the intellectual investment required to understand complex scientific evaluations and environmental decision methodologies than are participants in 'normal' political decision processes, judicial enquiries, or individual-based participatory forums. Furthermore the cross-group character of the deliberating body can allay public fears that partisan 'experts' are manipulating policy-making through their control of specific methodologies and techniques.

Sixth, co-operative management regimes can provide a framework which *encourages environmental learning* - enhancing social capacities to define and redefine, to solve and resolve, environmental problems. Studies of policy learning suggest that significant progress in the ability to manage problems typically occurs in relation to focused issue domains, where contacts among concerned actors are to some degree conflictual but where tensions are bounded by interdependencies and discursive interaction. Collaborative management structures can assure precisely such a context, facilitating interactive 'learning' as participant groups engage with each other within a structured framework in an iterative process of defining problems, identifying solutions and initiating practical reform. (Lee, 1989; Glasbergen, 1996).

There are other ways in which the potentials of such systems of environmental governance can be appreciated. In a sense they allow all participants to manage *risks* more effectively. Inputs from a wider range of actors at the stage of defining the problem should allow a more accurate cataloguing of relevant risk factors and risk perceptions. Actors from each of the three social sectors referred to above can gain on this front. Politicians can to some extent distance themselves from detailed policy-making, while basking in the beneficial results of an 'agreed' solution. Business can gain policy stability and ordered change. Environmental organisations can begin to have an impact on shaping practical policy outcomes. Furthermore, in terms of the features typically associated with environmental problems (Dryzek, 1987; Lafferty and Meadowcroft, 1996), their inter-connectedness, geographic dispersion and complex patterns of effected interests are accommodated by the flexibility of defining the initiating problem area and the range of social participants, while longer time frames can more easily be addressed in groups that are relatively insulated from the normal electoral cycle.

2.4. Objections to co-operative environmental management

There are many kinds of criticisms which could be levelled at the ideal-type environmental management model sketched out here. We shall consider four lines of argument which cover much of the ground.

First, there is *the argument from power*. At the heart of this objection is the observation that real world politics is not about negotiation among equals, but power-centred interaction. Thus the mechanisms described here are not really 'co-operative' at all: rather each party brings resources to the table, and the 'agreed' solution will ultimately reflect existing differentials of power. Furthermore, structural circumstances necessarily favour organisations of producers over consumers, business over labour, corporations over environmental pressure groups, and the rich over the poor. Thus, powerful actors like state agencies, bureaucracies, business federations and multinationals can be expected to dominate such bodies if they consent to take part, or to vitiate their activities if they stand aloof. This objection therefore focuses on the 'naive' or 'idealistic' resonance of co-operative management initiatives, and remains sceptical of the liberal pluralist assumptions behind the approach which do not appear to address seriously the structural inequalities and systemic imperatives which constrain reform.

Second, there is *the argument from democracy*. In addition to the 'anti-democratic' implications of the objection traced above (that the more powerful groups will always get their way), it can also be argued that most of the groups participating in these processes have weak democratic credentials: business organisations and even environmental action groups seldom function on democratic lines internally. So who do these groupings really 'represent'? Why should solutions negotiated with representatives of special interests be preferred to policies independently decided by democratically elected assemblies at local, regional or national levels. In fact, it could be argued that such 'group-based' processes inevitably undermine genuine democratic government both by transferring important decisions away from responsible officials into the hands of pressure group cartels, and by degrading the quality of public debate by emphasising parochial allegiances over the common good.

Third, there is the *argument from efficiency*. What guarantee is there that the outcomes that are 'agreed' really will address adequately the grave environmental problems they are intended to solve? Will not the negotiation process consume substantial resources of time and energy, in order to produce a 'lowest common denominator' policy which will soon reveal its inadequacy - discrediting both the process and the participants? Surely more progress can be made if each social sector concentrates upon what it does best. Certainly many businesses would prefer government to take its responsibilities in hand in a straightforward manner, regulating (or deregulating) to create a clear context in which operational decisions can be made by private parties. And perhaps environmentalists can contribute more to improving societal environmental performance by acting as watchdogs and mobilisors of public attention rather than by trying to participate in public decision-making functions which they are ill-equipped to fulfil.

Fourth there is *the argument from political culture*. The point here is that co-operative management may have some place in societies which display a long tradition of consensus-based decision-making - such as the Netherlands or perhaps Scandinavia. But patterns of patient negotiation, consultation and mutual accommodation are alien to the individualistic and liberal political cultures typical of countries such as Britain and the United States, and so co-operative management forms are unlikely to play any significant role there.

Each of these arguments raises serious issues, and taken together they might seem to undermine significantly any claims that could be made for co-operative environmental management. Yet the real question is not whether co-operative management has problems, drawbacks, and dangers, but whether in particular contexts it may achieve more satisfactory results than traditional regulatory approaches.

The argument from power
The central claim of the argument from *power* is that because co-operative management exercises fail to recognise power differentials and structural constraints, the outcomes will be skewed in favour of dominant interests and/or that talk of 'consensus' will serve as an ideological cover for a continued refusal to address seriously environmental problems. It is true that social actors command different sets of resources which they can deploy to influence social decision making. Furthermore, deep structural features of social and economic life constrain the options of individuals, groups, and society more generally. But these are realities which any proposed pattern of decision making must accommodate. Co-operative management addresses them in a variety of ways. Participant group selection must be based on existing patterns of interest, activity and organisation.

Powerful 'players' cannot be ignored. On the other hand, groups which have previously been excluded from closed policy-networks can be brought within the ambit of the process. Negotiations will necessarily take account not just of 'what is said', but also of 'who says it'. And yet the format of collective debate, investigation, and decision can be structured so that knowledge, evidence and argument count in deliberation. Again, the idea of the discursive reconstitution of interests, and the long term and iterative character of collaborative management processes, can encourage the gradual change (of otherwise inaccessible) structural circumstances.

The power argument might be taken to suggest that we should rely on government (which can enforce compliance) rather than 'negotiation'. But of course this is to assume that government already understands the problems, can and should decide how to distribute the burdens of adjustment, and is able to implement solutions without the co-operation of other social actors. If this is true then co-operative management is indeed unnecessary. But the experience of the last two decades suggests that in many circumstances it is not true. Often it is only through engagement with groups representing different aspects of an issue that the ecological and political dimensions of a problem can be adequately mapped; only in partnership that equitable and effective solutions can be identified; and only in conjunction with multiple actors can solutions be enacted.

Or, the power argument might be taken as advice to environmentalists to put their faith in building their movements, rather than becoming too closely tied to state and business interests. The fear here is really one of co-optation - that close contact with political authorities and business leaders will corrupt the authenticity of the environmental voice. Of course, it is only because of educational and campaigning activities that environmental groups have earned the popular recognition than can win them a place at the negotiating table. Yet if they do not take a direct part in making policy, and in collaborating with other social partners to implement solutions, the practical gain from their activity is limited. In fact, one alternative does not exclude the other. An environmental action group can maintain its mass-based activities, and yet participate in a relatively elite exercise such as co-operative management. Needless to say, it cannot *simultaneously* support *and* denounce (a particular process or outcome), but it can endorse agreement in one environmental area, while criticising progress in another. Furthermore, from the point of view of the 'environmental movement' as a whole it is likely that at any given time some groups will participate in co-operative management while others will steer clear of such arrangements.

Alternatively, the critique from power may be construed as a 'realist' counsel, that co-operative management will be able to do little to alter (environmentally destructive) behaviour which is rooted in fundamental characteristics of the existing political-economic system. Yet one of the accomplishments of contemporary social science is to have established that institutions can 'make a difference'. Relatively small shifts in institutional configurations can profoundly alter social and economic outcomes (Ostrom, 1990; Weimer, 1995). Co-operative management structures cannot be expected to set all the world to rights, but they can contribute to change.

The argument from democracy

The central concern of the argument from *democracy* is that the authority of elected institutions will be undermined by transferring environmental decision making to an array of management bodies controlled by special interests and pressure groups. This perspective can be understood as compatible with either a defence of existing 'representative' political mechanisms, or a plea for further democratisation based on increased citizen participation in the work of government and/or 'workplace democracy'. With respect to the first perspective, it should be noted that the framing conditions within which co-operative management regimes can operate would continue to be set by national political institutions. Furthermore, specific agreements would generally require some form of approval by local, regional and/or national political authorities. Moreover 'public' bodies (central government departments, regulatory agencies, local and regional governments) will themselves be partners in such arrangements. Thus the orientation and activities of such mechanisms would not escape the control of democratically elected governments. Indeed, since today the detailed work of environmental management is usually conducted by administrative agencies (often in liaison with a few major interests), and not directly by elected representatives, co-operative management can actually open the process to greater democratic scrutiny.

With respect to the second perspective, co-operative management does not necessarily preclude more direct citizen participation - for example, a local referendum to reject or approve implementation of a negotiated package. But co-operative management does privilege group inputs over individual-citizen inputs. It suggests that the cross section of interests and perspectives involved in environmental problems can best be represented, and conflicts resolved, by bringing together leaders of the diverse interest constituencies. This combines a plurality of inputs with authoritative representation - participants with time, expertise, and an organisational base - within a collaborative decision-making framework.

The argument from efficiency

The argument from *efficiency* involves at least three distinct points. The first focuses on the process costs implied by this sort of management regime. Are participation costs so high that groups will shy away from taking part? The second focuses on outcomes. What guarantees are there that co-operative regimes will achieve the intended environmental (and perhaps developmental) objectives? The third focuses on co-ordination across policy sectors. Will not the proliferation of co-operative management regimes lead to fragmentation, self-contradiction, or redundancy within the environmental regulatory domain? With respect to the first point, participant organisations will have to bear certain costs - at the minimum there will be those associated with liberating senior decision-makers to take part in investigation and face to face negotiation. Organisations may also be required to make a direct financial contribution - to cover the ongoing activities of the encounter group, pay support staff, commission collective research, and so on. In many circumstances government will have to provide the bulk of the support funding. Whatever the case, groups will assess direct costs within a wider context of the potential gains and losses they anticipate from the exercise. And this will encompass many factors including their estimate of the severity of the issue around which the collaborative mechanism is being drawn together, their assessment of alternative ways to influence outcomes, whether they believe the process can proceed without them, the possible impact of participation/non-participation on the organisational coherence of the group, and so on.

With respect to outcomes, environmentalists and government departments charged with protecting environmental resources may well be concerned that the inevitable upshot of negotiation processes with business will be to 'water down' standards, defer action, or stretch out the time frame over which environmental improvements are to be achieved. There can be no advance 'guarantee' that a negotiated solution will prove to have adequately addressed the issues which initiated the exercise. But certain factors would seem to increase the likelihood that this might be so. The initial selection of groups is obviously important, particularly with respect to whether the key sectors/interests/perspectives have been represented. The ways in which various kinds of knowledge (ecological, political, economic) are integrated into the decision process, and the ways in which risks and risk perceptions are addressed, are also crucial. Transparency of process is also significant - since public monitoring of agreements provides an independent check on the course of action determined by participants. Most important of all, however, is the establishment of agreed procedures for verifying impacts, and then reconvening to consider results at regular intervals. In this way issues which were not resolved in earlier rounds can be brought back to the table, unanticipated consequences can be appraised, and weaknesses in the original orientation rectified.

With respect to co-ordination across policy domains a substantial role will remain for central government in fixing the framing conditions within which co-operative exercises may operate. Yet if these guidelines are sufficiently flexible to allow groups of social partners to derive specific answers to individual problems, they are also almost certain to permit contradictory initiatives, duplications of effort, disjointed actions, and a 'patchwork quit' of programmes and environmental policies. It must be frankly acknowledged that fractured policy-making is to some extent an inevitable result of decentralised and negotiated problem-solving. Yet the alternative of some single, coherent and fully integrated policy response to environmental quandaries is illusory (Meadowcroft, 1997b). This is particularly true with respect to the implementation of sustainable development, where we are only just beginning to understand the scale and direction of the changes this will imply. In such a context redundancy and overlap, and a diversity of experiments with solutions, is especially desirable. Within a pluralist democratic system many social processes, including public debate and oversight, and 'partisan mutual adjustment' (Lindblom, 1979), will operate to counteract some of the negative effects of fragmentation. Furthermore, government agencies can themselves assume a surveillance mode - actively searching out management gaps and incoherencies, and encouraging fresh initiatives and negotiations to resolve them.

The argument from political culture

With respect to the argument from *political culture*, it is certainly true that there are great differences among the political and administrative traditions of the developed industrial states, as well as significant differences in business culture and the practises of pressure group organisation. It is no accident that the best known examples of co-operative management are to be found in the Netherlands, which has long served as an example of a consensus-based political system, which stands at the opposite end of the cultural continuum from the individualistic, confrontational and litigious culture of the United States. Observers of the American scene have in particular noted the difficulties presented by the fact that political bargaining (log-rolling) has acquired an unsavoury image (Rabe, 1988), and that (over) detailed legislative provision by Congress circumscribes the negotiative authority of regulatory administrators (Fiorino, 1995). Yet does this mean that co-operative management has no future outside consensus cultures? Not necessarily. For the pressures which are encouraging negotiated environmental settlements - including the complexity of environmental issues, the range of concerned interests, the uncertainty about future developments - are clearly manifest in societies that are habituated to less consensual policy making. Indeed, even in the United States - where the confrontational stance of industry and environmental representatives, the tendency of seeking 'all-or-nothing' solutions through the courts, and the detailed legislative frameworks which constrain regulatory flexibility, create substantial barriers to negotiated management regimes - steps are being taken in the direction of co-operative management. Consider, for example, the United States Environmental Protection Agency's recent 'Sustainable Industry Project' and the 'Common Sense Initiative', both of which emphasise co-operation between regulators, industry, and community and environmental groups in order to improve environmental management on an industry/sector basis (Fiorino , 1996). Above all, even in countries lacking a 'corporatist' or consensual decision culture, the marked and continued frustration with established regulatory mechanisms and policy is encouraging the search for new ways of doing things.

2.5. Conclusion

This chapter has sketched a model of environmental management that relies upon collaborative interaction among groups from different social sectors to develop and implement creative solutions to environmental problems. It has suggested potential advantages of such practices, and considered some of the more obvious objections to this approach.

Like other apparently favourable schemes to reform social administrative structures the difficulty lies 'with the details'. A wide range of variables may effect the practicability of such arrangements, and determine whether in any given instance they can achieve the positive impacts, and avoid the problems, considered above. The existing literature on inter-organisational collaboration, negotiated conflict resolution, and network management suggests that the viability and outcomes of cross-sectoral co-operative environmental management will be highly sensitive to factors such as: the structure of interest representation in the problem domain (what interests are present, how they are organised, and how power resources are distributed); the parties' perceptions of the risks and gains of collaboration and of the potential costs and benefits of pursuing alternative strategies; how the process is initiated and who selects participants; the way in which encounters are structured, in particular whether the interaction is 'managed' by a facilitator or mediator; the success with which participants' core concerns are accommodated and imaginative solution spaces opened up; and the range and character of anticipated implementation activities. These literatures also raise the difficulty of assessing the relative 'success' of such arrangements (Glasbergen, 1995). Should performance be judged on the basis of the attainment of the original ends of the primary instigator, implementation of the agreed accord, stability of the co-operative arrangement, or realisation of some specific degree of environmental protection? All these issues merit study in relation to environment-centred co-operative management initiatives. But three other themes are of particular importance.

The first relates to problem definition. Which sorts of problem domain offer the most promise for this form of management system? What sorts of procedures and techniques are best suited to various types of problem domain? How should group selection, encounter techniques, and implementation measures vary to accommodate contrasting problem domain types? How do the time scales over which exercises extend interact with these factors? Particularly important in the context of sustainable development implementation is the issue of the sorts of problem domain that allow environmental and development concerns to be most constructively juxtaposed. No doubt regionally focused initiatives are of particular promise here (Driessen and Vermeulen, 1995). After all, environment and development problems are often experienced most directly at the local level, the proximity between local publics and decision-makers favours participation and two-way feedback, and there may be a strong sense of (place-specific) identity - all of which can favour collaborative interaction.

A second theme relates to the range of social partners included in co-operative management initiatives. Contradictory pressures are at work here, for while a more inclusive process will ensure that a greater variety of interests are given voice and that a broader support base for a negotiated solution is generated, it may also add dimensions of complexity and conflict to the interactive process which precludes emergence of an outcome acceptable to all partners. Agencies initiating co-operative ventures must give careful consideration to identifying core groups without which a workable management strategy cannot be implemented. The most delicate issue typically concerns participants from the 'not-for profit'/NGO/civil society sector, which in organisational terms represents the weakest of the spheres from which co-operative partners can be drawn. Environmental action groups and community organisations lack the financial and technical resources of government and business participants. They usually represent more diffuse constituencies and have less experience with detailed policy-making. They may have cumbersome internal decision processes or lack a degree of professionalism. Furthermore, there may well be a number of rival organisations claiming to represent the 'environmental option' on a particular terrain. In many respects these groups represent a 'pain in the ass' for government and industry participants, and there is a strong temptation to proceed without them. After all, even when such groups take part in negotiations they may prove unable to enforce an agreed solution on their supporters, and even if they succeed at this, other more confrontational groups await in the wings to oppose any negotiated solution. Yet there are major advantages - both for individual management exercises, and for the development of this approach generally, that can accrue when participation from the civil society sector is secured. Environmental and community action groups bring distinctive perspectives to the negotiating table. They can present governmental and industry groups pre-occupied with immediate political and economic constraints with alternative perspectives, reminding them of broader obligations to the public, future generations and nature. They can help confer legitimacy on collaborative management initiatives in face of widespread public scepticism over the intentions of government and business. They can provide useful 'checks and balances' against any tendency for negotiative processes to lapse into cosy 'insider-group' accommodation, which fails to address the gravity of the environmental challenge. Participation in such processes can also expand the horizons of non governmental organisations, increasing their capacity to serve as collaborative partners in the future. Moreover, by skilfully integrating such groups into dialogic processes public sector agencies may be able to leverage more rapid and innovative solutions from implicated economic interests than might have been the case had regulator and business conferred behind closed doors. It should be remembered that co-operative management does not necessarily imply complete symmetry of participation, and different levels and degrees of integration into the process may be appropriate for different forms of organisation.

The third theme relates to political, legal and regulatory contexts, and to the sorts of changes to existing governmental practices which can favour the development (and positive social impact) of co-operative environmental management. Here questions focus on the sorts of reform which - in the context of the specific institutional frameworks and political cultural environments found in different industrialised societies - may unlock the potential of such management systems.

It has already been argued that co-operative procedures are not intended to replace the broad scale regulatory structures that have been built up to handle environmental difficulties in the developed countries over the past thirty years. Even if the co-operative model was adopted widely, this would not eliminate the need for 'clean air acts', environmental protection agencies, pollution inspectors, and so on. Indeed co-operative management is only likely to make inroads in a context where government has expressed a clear commitment to improving environmental standards, and to encouraging social institutions to engage with the challenge of 'sustainable development'. Otherwise, the continued influence of traditional economic interests in policy making, the (relative) weakness of environmental constituencies, and institutional inertia, would make it unlikely that groups would be motivated to invest in this type of collaborative venture. What is particularly important in this regard is the maintenance of a 'progressive environment dynamic' - the establishment of a clear impression that environmental problems are here to stay, that they must be addressed progressively, and that in the future the demands made by government (but also by citizens, investors, business partners) are likely to be more (rather than less) stringent. Above and beyond such a general posture, public authorities that desire to encourage such collaborative initiatives need to do several things. They need to establish core goals for particular areas of environmental policy (objectives, targets, limits): goals that are sufficiently stringent to oblige prominent actors to modify current behaviour, and yet sufficiently flexible to leave open major choices over how such general goals can be accomplished. Such a 'policy scissors' is designed to encourage groups to recognise that it is in their interest to collaborate with others to establish a management regime which, while operating within the general framework of government policy, would allow groups directly concerned to reached preferred solutions (Driessen and Vermeulen, 1995). Government could also act more directly as a facilitator - encouraging encounters between relevant organisations, legitimating such procedures, providing core funding, and so on. Governmental agencies at all levels would also need to be encouraged to participate as partners in the collaborative regimes, and the time of senior officials would have to be allocated to this end. Moreover legislative or administrative reforms might be required to permit negotiated solutions to be carried through in practice. Thus, while co-operative management is in one sense a step away from simple 'command and control' regulation, in another it requires an active initiatory role by governments.

Finally it should be observed that co-operative management is not predicated upon the assumption that social relations are essentially harmonious, that negotiated solutions can be found to each difficulty, or that every environmental conflict can be converted into a win/win scenario. On the contrary, social interaction is rife with conflict; political problem resolution often depends upon forcible closure and upon the imposition of discernible losses on one group; and agreements are always relative, partial and transitory - destined to be disturbed or overthrown by subsequent developments. Yet precisely because all this is true, it would seem important that we do our utmost to explore the potential that co-operative and negotiated interactions offer to tackling environmental issues, which offer some of the most complex challenges facing contemporary societies.

3. ENVIRONMENTAL GOVERNANCE AND MODERN MANAGEMENT PARADIGMS IN GOVERNMENT AND PRIVATE INDUSTRY

HUIB ERNSTE

3.1 Introduction

This book is an inquiry into the nature and preconditions of public-private co-operation for managing environmental problems. That topic calls for an investigation of the place of such co-operation in public administration as well as within the private sector. In both realms, one can observe a continuous search for innovative managerial strategies. Several modes of environmental governance were already introduced by Pieter Glasbergen in the introductory chapter to this volume. Also on the side of private industry, various organizational paradigms have evolved over time, some becoming more dominant than others. Until a few years ago, organizational paradigms for the public and the private sphere were developed independently. The new trend towards 'New Public Management', on the other hand, draws heavily on experience in private industry. The ties are especially clear with what is called the 'Lean Production' paradigm, creating substantial parallels to the main strategic thrust of these modes of organization. In this chapter, I will investigate the extent to which this common ground is a resource for public-private co-operation in the field of environmental management.

The ultimate aim of both Lean Production and New Public Management is a more efficient and effective organization. To improve the efficiency and the effectiveness of their organizations, entrepreneurs have three options, irrespective of their context managing a policy programme or a private firm. The alternatives are: 1) market, 2) hierarchy, and 3) co-operation. The choice is mainly a question of which social relationship between the different actors and people concerned has the greatest potential for a rational improvement of the processes involved. Most organizational paradigms consist of a specific combination of all three modes for co-ordinating the actions of all relevant actors involved in the process of production or policy implementation. In our case, the main question is: 'What role does co-operation play in schemes such as New Public Management and Lean Production?'

P. Glasbergen (ed.), Co-operative Environmental Governance, 43–63.
© 1998 *Kluwer Academic Publishers. Printed in the Netherlands.*

Listening to the rather uncritical and ideologically biased slogans of those 'management gurus' trying to sell their ideas on an increasingly contested market, we get the impression that the main thrust of the New Public Management concept is the introduction of internal and external markets for public services and their intermediary products, as well as the promotion of entrepreneurial behaviour among state actors on all levels. However, if we trace the roots of these concepts, the picture looks quite different. In this chapter, I argue that it is not so much the introduction of market principles but rather both internal and external co-operation including public-private co-operation that constitute the innovative core of these new management paradigms. This is illustrated by Swiss examples.

3.2 Developments in environmental policy implementation

Concepts for effective and efficient environmental policies do not drop out of nowhere; rather, they are tied to a historical context, which develops differently in each country. Despite this contingency, a pattern of stages can be recognized. Within this common pattern, several lines of development can be distinguished (Glasbergen, 1994). First of all, the mix of instruments for environmental policy has changed. Second, the perception of environmental problems has altered. And third, the organizational structure of the state agencies responsible for the implementation of environmental policies seems subject to a fundamental re-structuring. In this chapter, I will mainly address the third line of development.

Stage I: direct steering
Early environmental policy was mainly a fire-fighting policy, so to speak. It sought to protect the population from threats to their health. Environmental policies at this stage were primarily technical, end-of-pipe solutions. The most important instruments for environmental policy-making in those days were licences and permits. This is a typical case of a command-and-control strategy, which enforces certain standards top-down by means of regulation, controlling, and sanctioning. This is the first ideal-typical model for environmental management, namely the direct steering model.

The direct steering model is based on the idea of a society as a technical system which can be directed at will by the state decision-makers. The state is the only omnipotent steering agency of this societal machinery. It defines the operational goals of environmental policies which it deduces from scientific knowledge. The executive state agencies then design the concrete measures to be taken, which are supposed to directly influence the actions of the individuals and private organizations. These therefore lack the possibility to express their preferences or to evade the decisions made centrally.

This model corresponds to the instrumental ideal of a logical, consistent, effective, efficient, and predictable society. That society is comparable to a machine of which the blueprints and the functioning are exactly known and steerable (March & Simon, 1976, p. 39; Kieser & Kubicek, 1978, p. 78ff.). If the society does not react in the desired way, then these tendencies call for direct counter measures to restore a sustainable society free of surprises. If this is not feasible, the therapy is self-evident: deepening the natural scientific or technical knowledge of ecological causalities, tightening the standards, sharpening the sanctions, and imposing strict controls. It is only logical that requirements and prohibitions are preferred within this framework. Once these are issued, those concerned can only obey or practice civil disobedience.

When this command-and-control policy diffused into an increasing number of political fields, its limitations became apparent. It created a dense undergrowth of regulations, in which nobody could find his way anymore, paralyzing bureaucratic procedures until a deadlock was reached (Mayntz, 1978; Hennis, Kielmannsegg & Matz, 1979). In Switzerland, for example, it is commonly observed that exception permits are issued over and over again, allowing firms to pollute more than the official standard. By trial and error we had to learn the hard way that an environmental policy based on legal measures is not enforceable at the push of a button. On the contrary, the state is not always able (or willing) to enforce its own environmental regulations. Then too, sanctions alone were not enough to make parties comply with the regulations, if they were not inclined to do so. A whole series of scandals showed that intrinsic motivation was often lacking.

Stage II: indirect steering
First, an attempt was made to grasp the complexity of the problems and to identify all relevant actors involved. This was done by redefining environmental problems as causal chains and by introducing a process orientation. Second, the failure of traditional environmental policy led to the reconsideration of the division of labour between state and society. The state alone was not able to solve the problems. More space was needed for initiatives by non-governmental organizations and private actors. Instead of directly influencing the actions of specific actors, the state had to confine itself to the creation of conditions which allowed the actors to take certain initiatives themselves. This pragmatic argument was joined with the neo-classical tendencies of the Reagan and Thatcher era. The image of a state taking care of its citizens from the cradle to the grave was replaced by a solicitous society based on independent, rationally calculating, egoistic, and exclusively purposively acting individuals. It was recognized that

increasing integration and concentration on core tasks parallel to a streamlining effort (if not to say 'deregulation') was necessary. Third, in the quest for new and more effective environmental policies, the instruments used were diversified. Increasingly, market instruments became the focus for the indirect steering of environmental action. This paves the way for the second environmental management model.

The economic steering model is based on the idea of a society as a large market with many autonomous and competent actors directed by an invisible hand. Not the state, but the participants in the market know what kind of environment they prefer and how they can best fulfil their aims. Presumably, they will maximize their own utility by choosing the most efficient means to achieve their goals. The market then co-ordinates these activities in such a way that a (Pareto-)optimal result will be obtained. The state only interferes indirectly. It ensures that a market exists for environmental goods, and that free market principles are working. The main goal of environmental policies in this context is the internalization of external effects. The valuation of the environmental goods and the prices people are willing to pay for them is incumbent on the market actors. They decide if and how to react to price signals and offers and how to behave on the market. The definitive impact of an environmental policy in such a situation is not determined by the state but implicitly by individual actors in the market. It is the market which finally determines how much pollution is acceptable and which environmental goal is reached. Market instruments take these preferences for granted and are not designed to change them. It is one of their limitations that they only secure efficiency of means and not goals.

Still, it is clear that the enrichment of the environmental policy spectrum with market instruments led away from a pure environmental repair and protection policy, consisting mainly of end-of-pipe solutions. The new direction was an avoidance policy, corresponding to the precaution principle of environmental policy-making.

Stage III: self-steering
In the last few years it has become clear that indirect steering via the market has its limits. The necessary free market conditions only exist in rare cases. Many politically powerful actors have an interest in avoiding the introduction of free market conditions according to which they would have to pay the full cost of the pollution they produce. Often, they would rather lay the responsibility for a sound environment on the state (Goverde, 1993, p. 77). We should therefore not get carried away by free market illusions. Like the command-and-control system, market instruments also depend on the acceptance and legitimation of these instruments by the institutions and individuals concerned. As a consequence, more

pragmatic and ideologically less biased environmental policy-makers have turned to another environmental model, namely the negotiation and co-operation model or the self-steering model.

The consensus or 'self'-steering model is not based on purposively directing the actions of the parties involved but rather on the initiation and maintenance of a continuous (societal) learning process. In analogy to operational processes of change in private industry, one could speak of a continuous social rationalization process (see also Reichert, Zierhofer et al., 1993, pp. 33-43). Rationalization in the indirect steering model relies on the exchange of environmental goods on the market. Therefore, it is confined to economic rationalization based on predetermined preference structures. The consensual steering model enhances this economic process with the rationalization of the goals, preferences, and valuations of the actors. The mechanism responsible for this kind of rationalization is not authority or force (as with the command-and-control model) or incentives (as with the economic model) but rather persuasion. In terms of modern social theory, this would be called 'communicative rationalization' (Habermas, 1987a, b). The state can influence such a social learning process, though not so much by taking a specific position itself and making this position plausible. Rather, the state can act as an initiator and moderator of a discourse between all involved parties. In doing so, the state would facilitate a decentralized social learning process. The role of the state is then confined to the initiation and support of a great number of co-operative processes involving all relevant social forces and movements. In this way, it can additionally mobilize basic social energies and innovative powers. Instead of pre-selecting which goals and solutions are to be realized, these goals and solutions are first developed and subsequently realized within the framework of these communicative learning processes. This model relies explicitly on self-regulation and self-determination. The relevant actors themselves (including the state) take the responsibility for a clean environment. The main instrument for this consensual steering model is the initiation and maintenance of co-operative networks. This steering model forms the basis for the co-operation principle in environmental policies.

As mentioned before, these developments in environmental policy implementation also have important consequences on the way government is organized. This is manifest in what is called the concept of New Public Management. I now return to the roots of this concept.

3.3 Lean Production

New Public Management is the magic word for a whole bundle of innovations in the organization of the state. The concept of New Public Management adopts significant elements from the Lean Production concept developed in the manufacturing industry. On the one hand, this parallel is not contrived. Both private manufacturers and the state have to deal with similar developments in society and in the economy. More specifically, it can be shown that the specific conditions which gave rise to the development of these new organizational paradigms in Japanese manufacturing industry are comparable with the conditions of many state organizations. On the other hand, the New Public Management concept has been deviating from the original central ideas of the lean production strategy. These differences are especially relevant to the role of public-private co-operation in the context of New Public Management. In this section, I will briefly recall the original ideas of the Lean Production concept as described in the groundbreaking book on the second revolution in the automotive industry[1] by Womack, Jones & Roos (1992, pp. 53-78).

With Henry Ford's fundamental innovations in the organization of mass production in the first half of the 1920s, Fordism emerged as a general organizational paradigm for manufacturing consumer goods. Productivity growth was sought by increasing the vertical integration of the production process. This was accompanied by a deepening social division of labour and further specialization by separating manual tasks from brain work (Taylorism) and by building up a huge bureaucracy to manage the whole process. Producing a long series of the same product led to enormous productivity increases. Changes in demand or in consumers' taste had a disturbing influence on this process. To ensure that long series could be produced, and to take full advantage of the increases in productivity, the production organization was more or less isolated from any potential disturbances, internal or external. This was done by stock keeping and creating special departments for the reworking and repairing of parts, half-products, and end-products. But the strategy also involved 'dressing' the corporate 'environment' through advertising, marketing, and public relations management. Buffers were also created between the different management departments and levels. To make this whole system manageable, it was necessary for middle management to filter all information it passed on to higher management levels. If changes in demand or other disturbing influences became too large to adapt to, production facilities were abandoned and employees laid off.

---

[1] Here I cite my own translation of the German title of the book. The original English title was: 'The Machine that Changed the World'. The title of the German version, in my view, gives a better picture of the contents and main message of the book.

The company then proceeded to construct a totally new facility adapted to the new circumstances.

In the same way, by the end of 1949, a sales slump caused by a general recession forced the Toyota Motor Company in Japan to dismiss almost one-quarter of its employees. This provoked a long and severe strike which ended in a historical compromise. The president of Toyota had to resign, taking full responsibility for the failure. Corresponding to the original plans, almost 25% of the work force was discharged. The others, however, received a life-long employment guarantee a nd regular profit shares. In this way, the employees became an integral part of the company community and shared in its destiny. From then on, they were a fixed asset which could not be written off and scrapped like a machine. Toyota was forced to re-invent manufacturing, taking these new circumstances into account. This was the birthplace of *Lean Production*. Since employees were now a fixed asset, the management had to make the best of this production factor. Flexibility and competitiveness had previously been achieved at the expense of the personnel, which had been expendable. Now, management had to achieve these goals while preserving its work force. This was possible by making a fundamental change in the organization of the production process, all the way from product development to selling the product to the customers. In this way, the industry could be inherently flexible and internally renew its competitiveness over and over again.

Integrative process orientation

In traditional Fordist organizations, product development usually started with a small development team and with a specific idea, which was handed down from top management. Later on, these development teams expanded to include other specialized teams working on specific development problems. Because of this expansion, of course, new ideas and new opinions emerged, but also and unforeseen problems. Management often had to backtrack and redesign certain parts that had already been developed. Otherwise, they had to accept post-hoc compromises to avoid further conflicts. In the new factory, research and development are organized in such a way that all relevant internal *and* external actors (such as suppliers and customers) were integrated into the process from the very beginning. These actors were expected to contribute their experiences and ideas but also express their needs. Therefore, companies started off with large project teams (instead of small ones). They negotiated all necessary issues until a compromise was found and an integral concept of the new product or production process was decided on (consensus by conflict). Once these priorities and goals were set, smaller teams simultaneously started to develop the different parts. This led to a vast improvement in the quality of the design and a drastic reduction in the total development and implementation time. Especially at the beginning of this development process, a major investment of time and resources was required of all parties involved, which entailed great risks. This new concept of product development could only be realized on the basis of long-term, open, and trusting relationships among the actors.

A lean, flexible and versatile organization

Another revolution took place in the organization of production itself. Instead of inflexible, highly differentiated, and vertically integrated firms, an effort was made to create lean, flexible, and versatile organizations. This was done by re-integrating many functions and responsibilities so that the workers in the primary value adding process on the shop floor would have much more influence on the production process itself. This increased motivation and quality. In this way, quality was not tested but produced. The hitherto largely neglected know-how and innovativeness of the workers on the shop floor was rediscovered and utilized. As a consequence, repair activities (the secret factory) -which in Fordist times comprised up to one-quarter of the total production capacity- could be avoided. Also, middle management could be reduced substantially and hierarchies flattened, since many managerial tasks were decentralized to the shop floor again. The surviving managers had to redefine their functions as facilitators, catalysators, promoters, coaches, mediators, or organization developers. Continuous improvement of products and processes now became the main goal of production, thereby boosting quality, innovation, and flexibility. A critical review of material and non-material inputs resulted in an extremely parsimonious and frugal production process. This applied to all stages, from raw material to final waste disposal, based on a low level of resource inputs. Information flows were not hierarchically organized but directly linked to the actual production processes, so that time buffers, large stocks etc. could also be drastically reduced. Reciprocal dependencies were not 'covered up' by these kinds of filters and buffers. Rather, they were explicitly given priority to be optimized, resulting in what is called a just-in-time production system. The main purpose of every production organization is, of course, the co-ordination of all the steps and parts of the whole production process irrespective of whether that step was actually being carried out by an external or internal supplier. In a versatile market environment, and with the built-in rigidities of the Fordist system, vertical integration of the whole production process within one huge centrally led and bureaucratically organized company was not feasible anymore. Many parts were therefore contracted out. Toyota often kept part of their capital resources in the newly created sub-contracting firms, assisting them with credit for necessary equipment. They regularly exchanged personnel and organized 'supplier clubs' to improve the exchange of information and co-operative problemsolving. To create such an agile but sensitive production system, the relationships between the different departments and production steps, as well as the relationships with external suppliers and customers, had to be based on long-term co-operation, reciprocal responsibility, and trust. Suppliers and purchasers did not build their relationship with the company upon the principles of power and counterpower but on the basis of shared destiny. Reciprocal transparency (e.g., *open book costing*) and truthfulness are the preconditions for such a relationship. This contrasts sharply with the usual attitude of rivalry, cut-throat competition, and the exclusive, self-interested, belligerent, negotiation mentality aiming at short-term benefits (Glasl, 1993, p. 119; see also Carlisle & Parker, 1989).

A customer orientation
Finally, the Lean Production system would have been senseless if it had not produced what the customers demanded. In this respect, Toyota did not go the usual way. Instead, they put together an extensive network of dealers who maintained close, long-term contacts with established regular customers. Their satisfaction with products and their suggestions for change were continuously monitored and fed back to the product development teams. The same principle is currently being adopted by many firms in the form of client-relation managers. Sometimes clients or customers are even organized in 'customer' or 'user clubs'. This kind of customer orientation is, of course, much more than just a defensive PR strategy, well known from former organizational concepts. The new kind of customer relationship turns information exchange, joint problem-solving, and decision-making into key strategic tasks. If, for example, certain problems might arise in the delivery process or the quality of the delivered goods might not meet the specifications initially agreed on, this would no longer lead to sanctions or compensation claims but rather to a joint problem-solving procedure. The same kind of customer orientation was also established for internal customers at the interface between different production units.

Internal integration and external association
In general, one can identify six basic principles of the Lean Production paradigm. These are listed in Table 1. In the context of this chapter, the first principle deserves extra attention. It can be further subdivided in two sub-principles (Glasl, 1993, pp. 99-103). One is *internal integration*; the other is *external association*. The second notion is a logical extension of the first.

Table 1: *Basic principles of Lean Production*

1) Integrative process orientation including external actors
2) Communicative and co-operative management
3) Mobilization of workers' know-how and innovatiness on the shop floor
4) Continuous development in small steps
5) Thrifty procession of resources
6) Client/Customer orientation

Internal integration. By goal and customer orientation, as well as strict process orientation and giving priority to bottom-up self-organization instead of top-down external control, the highly differentiated and specialized functions of the former vertically integrated Fordist firms are being re-integrated again. By doing so, a maximum proportion of employees can take full and better integral responsibility for the main production processes. This releases a broad innovation and optimization potential. As a consequence of directly integrating all internal functions in one single process or value added chain, clear-cut interfaces were created, with the suppliers, sub-contractors, capital, and know-how owners on the one side and dealers and customers on the other side.

External association. Since material flows and functions like product development always transcend the borders of the organization, external association is a logical consequence of the re-integration process. The external association feature describes the integration of important actors outside the company. The company integrates these external actors and interest groups by developing future-oriented, pro-active, collaborative, learning relationships. In this way, not only the internal value chain, but also the external value chain can be optimized.

In total, the industrial engineers of Toyota Motor Company needed almost twenty years to bring the Lean Production system to perfection. It was only in the 1970s and 1980s that it diffused around the world, where it still dominates the modern manufacturing industry (Piore & Sabel, 1984).[2]

The concept of New Public Management is an implementation of the same basic ideas in public administration.

---

[2] See also 'Wissen Spezial' of *die Zeit*, No. 14, 28 March 1997.

3.4 New Public Management

From the description of the Lean Production concept in the foregoing section, we already have a first glimpse of the concept of New Public Management. Originally, that concept was an attempt to apply the basic ideas of Lean Production to public administration. Some examples from Switzerland are described below. A review of the main principles of New Public Management are reproduced in Table 2 (Metzen, 1994, p. 125).

'Energy Model Switzerland' and 'Energy Model Zurich'
New Public Management demands 'lean laws' which go no further than formulating the final objectives (output, impact, outcome). The detailed technical regulation necessary to reach these goals is left to the relevant agents in the field. An example of such a lean law is the energy law of the canton of Zurich. It defined the general overall energy goals and allowed large energy consumers[3] to create pools in which they co-operatively operationalize these goals and design and implement measures. At the same time, they would be relieved of all detailed prescriptions and regulations in the field of energy use. The Energy Model Switzerland is a similar management-oriented approach invoked by the Swiss federal government to support these kinds of co-operative processes within Swiss manufacturing industry. Both models are based on a step-by-step social learning process, relying on a responsible co-operative discourse between the companies within a region or industry to agree on binding energy-saving objectives. These self-regulating processes are moderated by a trained mediator[4] with the support of the National Energy Policy Programme 'Energy 2000'. At the same time, the use of energy is assessed and related to important key values. Additionally, a benchmarking process is initiated, resulting in indicators for future management objectives.

The first to start such a process were eleven big hotels in the Zurich area, the Zurich Convention Centre, and the Zurich Exhibition Hall, all associated in what is called the 'Convention Pool'. Each member of this group finally agreed in 1997 to reduce energy consumption by 1.5% per annum for heating and 2.0% for electricity per service unit produced for the next 20 years. This equals about 20,048 Kwh per year. Over the whole period, this implies a reduction of 27%. Since the administration of the Canton of Zurich itself is also a big energy consumer, they did not just want to motivate others to reduce their energy

---

[3] Large energy consumers are those who consume an equivalent of 500,000 litre of oil or more, which corresponds to the consumption of about 200 households.
[4] Lukas Herzog of Alteno AG, Basle.

consumption but wanted to lead the way by agreeing to fulfil the objectives of the federal 'Energy 2000' programme. Several other associations of private actors have been formed and are now in the middle of the process, e.g.:
- *Lucerne Manufacturing Association (LIV);*
- *Association of Cement, Paper, and Cardboard Manufacturers;*
- *The plastics industry;*
- *Regional Association Ticino.*

Source: Graf, E.O., 1996; Evaluation des Energie-Modells Schwiez, Bundesamt für Energiewirtschaft Berne.
Media orientation on Convention Pool Agreement, June 24, 1997.

Advisory Commission for Electric Power Policy in the Canton of Berne
In 1997, the cantonal government in Berne created an advisory commission with representatives from politics, private industrial associations, science, and administration to work out new sources of energy to replace the electricity generated by the nuclear power station Mühleberg, whose licence runs out in the year 2002. The public administrators recognized that meeting their objectives would largely depend on substantial support in know-how and political acceptance by other agents in the field. A political mandate itself is far from sufficient in this critical context. The advisory commission consisted of representatives from politics, private industrial associations, labour unions, the Swiss Energy Foundation, the nuclear power opponents, energy experts, scientists, and representatives of the federal and cantonal government. The process is still going on. In March 1997, the working group presented six options which have since been worked out in more detail: (1) new natural gas power stations, (2) a new nuclear power station, (3) expansion of the hydro-electric power station Grimsel West, (4) importing electric power from abroad, (5) management of energy demand and increasing capacities of alternative energy sources, and (6) combining natural gas and steam power stations with decentralized heat pumps.

Source: Press bulletin of the Office for Information of the Canton of Berne, March 10, 1997; Personal communication by Project Coordinator Bernd Kiefer, Spring 1997.

Within the framework of the concept of new public management, the state is not seen as the only possible contributor to the solution of the specific societal problems the government is supposed to tackle. The state is just one of many actors in that specific arena. By means of policy networks, strategic alliances can be formed. As a consequence, the sovereignty of the state is partly given up and replaced by the principle of co-operation. Civilians and other actors, like non-governmental organizations and private firms, are taken seriously as partners.

They can participate in planning and implementation, especially in areas where market mechanisms are sure to fail.

Table 2: Basic principles of New Public Management

1. Group, team, network. All tasks are dealt with within the group. Problems are consensually solved. Internal competition is to be avoided. Everyone is a member of several horizontal and vertical groups.
2. Self-responsibility, autonomy. Everyone is fully responsible for their own activities. Each of these activities is based on specific agreements on goals, guidelines, and standards. If the demanded quality cannot be met, the activities are stopped and assistance is sought.
3. Feedback, information. All activities are accompanied by exceptionally intensive feedback. These reactions to the outside world serve the improvement and adaptations of the company's own activities. Information exchange assures reciprocal understanding.
4. Customer and market orientation. All activities are customer oriented. The wishes of the (internal or external) customers have highest priority. All relationships within the value chain are constituted as supplier-customer relations.
5. Priority for value adding processes. Value adding processes have highest priority within the state administration. This is true for the use of all available resources. Superiors and experts mainly work where the value adding processes take place.
6. Standardization. Formalization and standardization of working methods require the written and graphical description of all methods and standards. This implies the application of standardized methods for fast and secure communication.
7. Continuous improvement. The continuing improvement of performance dominates the organizational culture. There are no final quality goals, only steps towards optimization and innovation. The process is the goal.
8. Direct correction of failures at the root. Failures are seen as chances for process improvement. Every failure is seen as an uncontrolled disturbance of the process, an incident which is to be traced back to its original causes. If necessary, the firm will go a long way to find the causes of failures.
9. Pro-active thinking and planning in advance. The problems of tomorrow are sought out today. Not the successful reaction to failures but the avoidance of future failures by planning is regarded as an exceptional achievement (crisis avoidance instead of crisis management).
10. Small controllable steps. Development is achieved by taking small controllable steps. The feedback to each step directs the following one. The speed of innovation is determined by a rapid succession of steps, not by risky jumps.
11. Goal-setting and decision-making in consecutive cycles. Organizational goals are developed in consecutive and overlapping cycles, each covering several organizational levels (policy deployment).
12. Shop-floor orientation. Managers focus on the process taking place on the shop floor (in Japanese: '*genba*'). Within this framework, they concentrate on support and development of the human potential and on improving control over basic processes.

In general, neither modern private manufacturing organizations nor the state tend to isolate themselves from their environment. Instead, they integrate and adapt to that environment in collaboration with internal and external partners. Not only do suppliers and customers play a role, but the whole physical and social surroundings -including state agencies, corporatist organizations and private organizations- is involved. This is the domain of policy networks, negotiated solutions, voluntary agreements, joint action planning, and execution (Ernste & Sirol, 1995; Knöpfel, 1994) as well as consensual conflict resolution (Glasbergen, 1995; Claus & Wiedemann, 1994).

3.5 The difference it could make

New management strategies have turned many things upside down. Probably the most fundamental change was to halt or even reverse the process of specialization and alienation tendencies so typical of Fordist corporations and bureaucratic administrations. Instead of a continuing differentiation of specialized functions, which are managed by large bureaucracies wielding power and authority, functions have been increasingly re-integrated and workers in the primary processes have been empowered again. At the same time, many relationships which formerly were organized on the basis of harsh competition in a free but anonymous market are now co-ordinated within a co-operative network grounded in longterm trust relations. This is not just a marginal change; it is a rather basic one. Both Lean Production and New Public Management are organizational concepts. They are used to increase the efficiency and effectiveness of organizations by restructuring the organizations and by activating human resources. Both concepts focus on the same facets of the organizational structure: the humanization of working conditions; efficient processes; communication; co-operation; conflict resolution between persons, groups, and hierarchical levels; autonomy; participation; self-fulfilment; the experience of meaning and meaningful action; action identification; satisfaction; motivation; performance; scope; flexibility; development perspectives; individual values; and corporate culture. All of these facets of the organizational structure can be turned into goals for strategic management (Schäfer, 1996, p. 519). But why would all this really make such a difference? This begs the social-scientific question of the relationship between these strategic goals and the performance of organizations. The literature on these issues is dominated by management experts, engineers, and organizational consultants. On the one hand, this literature provides useful insights for practical work in this field. But on the other hand, it lacks a serious scientific elaboration of the basic issues at stake. Instead, it mainly builds on lay speculation and generalised expert opinions, or at best on analogizing scientific interpretation. One needs convincing arguments with a solid footing in social-science research for the envisioned change. Those arguments should be derived from clear-cut theoretical models, instead of everyday psychology and management philosophy (Schäfer, 1996, p. 520). In the context of Lean Production and New Public Management, this basic theoretical scientific reflection is of particular importance, as the original concepts are not immune to misinterpretation and distorted application.

A typical example of an interpretation which clearly diverges from the original core ideas is the adaptation of Lean Production in being crafted from Central European management practices. In contrast to the Japanese practice, Central and Eastern European companies interpret Lean Production -and more extremely so, New Public Management- not as a process of continuous small steps toward organizational development and maturation but rather as a forced and concentrated intervention, accompanied by drastic organizational, social, and personal changes. It is questionable, however, if it is possible to combine the decentralization of responsibility, the fostering of internal and external co-operation, and the reliance on innovative contributions of motivated workers with financial cuts, reduction of personnel, increasing internal and external competition, wage cuts, and limitation of career development potentials.

A theoretical reflection on the basic conditions for the functioning and success of the core concepts might instead clarify what is basically right or wrong with such interpretations and implementations. It would go beyond the scope of this book to delve into these more theoretical issues. The interested reader is advised to see Ernste (1998) for further elaborations on this issue.

3.6 Co-operative and communicative networks

According to the original New Public Management concept, the enhanced rationalization or learning process is supposed to take effect at two points. One is the interface between different phases of the value adding process, thus between internal and external organizational units. The other is between the members of each unit. In these arenas for communicative rationalization, (abstract) goals, concrete guidelines and specifications, (quality)standards, improvements, and innovations are openly discussed. The results are embodied in formal or informal voluntary agreements or joint plans. In this way, both administrative and business organizations truly become places of continuous political debate. And government and business are brought back to their place in society again. It is within these arenas of rationalization that the debate takes place on the definition and main action plan of sustainable development, combining the ecological, social, and economic needs and know-how of all stakeholders and relevant actors. These co-operative and communicative networks are supposed to provide the guidelines for future development and action within a highly flexible and dynamic society.

Modern business organizations and political administrations can only be flexible and efficient when they are part of a network of internal and external communicative processes. Their networks provide them with information, feedback, innovative ideas, know-how, resources, support, and directives for future action. In contrast to business organizations, the traditional political administrative systems derive this kind of directives and support from politicians. Furthermore, that input is supposed to represent the starting point for policy implementation. However, this separation between political strategic and technical operative management does not hold up in practice, especially with issues like environmental policies. The question of 'how'? is almost as political as the question of 'what'?. As a consequence, at every stage of the policy cycle, a 'political' debate has to take place in order to ensure the political legitimacy. Obviously, this can be done by returning the question to the politicians. On the other hand, it may also be done by continuously ensuring the consent of all the internal and external actors concerned at every stage, just like modern flexible business organizations do it. Direct democracy can be of some help here but certainly does not suffice. One has to pull out all the stops of participative (discursive) decision-making and of the co-operative and communicative strategy.

Much of this is already common practice within the framework of New Public Management. Typical instruments for assessing and continuously improving and evaluating policy programmes[5] within the framework of the New Public Management scheme are again supplier clubs, development teams, quality circles, user clubs, and customer surveys. Most of these instruments, however, are still only used within the field of the own organization and its direct suppliers and customers. It would be a logical extension of the external association principle (see the section on the origins of Lean Production above) to apply similar instruments in the field of environmental management.[6]

Of course, no real case is perfect. Practical cases of Lean Production or New Public Management can be positioned on a continuum. At one extreme we find what Werner Sengenberger and Frank Pyke (1992, p. 11) called the 'high road' and at the other extreme the 'low road' to industrial (and administrative)

---

[5] Aulinger (1996, pp. 269-279) speaks of normative management.

[6] A pilot project on regional environmental quality circles has been suggested by Ernste and Sirol (1995).

restructuring. The low road consists of seeking competitiveness through low costs and a deregulated market environment. 'Institutions and rules aimed at regulating competition are mere straightjackets, and should be kept to a minimum. [...] The problem with this approach is that the improvement it yields for competition, if there is one at all, is frequently short-lived. Mostly, in fact, it accentuates the *malaise*.' The increased pressure on costs, productivity, and other terms of employment and co-operation hinders the organization in acquiring and keeping the qualified labour and in maintaining its quality of services, products, and supplies, required for efficiency, flexibility, and innovativeness. 'So, in the absence of better performance and alternative possibilities, further cost-cutting may become inevitable, resulting in a vicious, downward-spiralling cycle.' In a sense, this is also what Blowers (chapter 11 in this book) tries to warn us about. At the other extreme -the 'high road'- competition is seen as constructive (in the sense of benchmarking). It comes close to the ideal form of Lean Production or New Public Management.

With this normative theory in mind, we can now take a closer look at the practical implementation of New Public Management schemes in Switzerland. In doing so, we focus on the role that private-public co-operation plays within them.

3.7 Effect-oriented administration: new public management in Switzerland

At first sight, Switzerland would seem to be the perfect crucible for New Public Management. Public administration in Switzerland was always known to be rather lean, efficient, parsimonious, pragmatic, and uncorrupted. The traditional decentralized federal structures seem particularly suitable for New Public Management. Also, the boundaries between the public and the private sphere have always been rather diffuse. The delegation of public tasks to private organizations and the many mixed non-parliamentary commissions[7] have a long tradition in Switzerland. This particular system is also known as 'militia government'. Nevertheless, there are reasons for scepticism. The political scientist Raimund E. Germann (1995), for example, mentions that the typical profile of these part-time honorary public functionaries contrasts sharply with the profile of highly professional (network) managers required by the New Public Management strategy. Most governments on the cantonal and federal level are 'collegial governments' without a specific leader. This encourages strong departmental particularism. Swiss governments are mostly large coalitions.

---

[7] Germann (1995, p. 84) mentions 350 of them at the federal level and more than 2000 in the cantons.

Consensus is often not much more than a proportional distribution of ministerial chairs. Still, in Switzerland -inspired by the success of the principles of Lean Production in private industry- a large number of New Public Management projects have been launched under the title of 'Effect-Oriented Administration'.

Yet, these projects have been carried out in a way which raises suspicions that something essential is being left out. As we have seen, Lean Production is -in theory- actually a major shift away from a system based on hierarchical and free-market co-ordination of human activities to a system of co-ordination by internal and external co-operative networks. In this sense, it is astonishing to note that the Lean Production concept advocated by the Effect-Oriented Administration scheme in Switzerland -and probably not just there- points in a totally different direction. It makes market co-ordination the core of its approach.

This is a fundamental misinterpretation of the original basic ideas and success factors of Lean Production and New Public Management. And it is one of the reasons why many of these projects in Switzerland will fail miserably. Specifically, it is the networking, the communicative and participative management, so basic to the Lean Production paradigm, which is -deliberately or not- dropped and substituted by some kind of quasi-market mechanism based on competition instead of trust. Let me explain.

The prospect of a lean state, contracting out (privatizing) certain tasks which formerly were executed by the state, rallied liberal and neo-liberal political forces. They have always sought to curtail the state to make more room for free-market forces as a co-ordinative mechanism. They did not seem to be bothered by the fact that Lean Production is not exactly a scheme which reaches this goal by introducing more market; it operates by a third, totally different co-ordinative mechanism, namely networks and communicative management. Still, at crucial points, the original ideas of Lean Production were altered (or misinterpreted). Those critical points were the interfaces between different stages of political goal formulation, via decentralized output providers to customers, where the Lean Production strategy originally foresaw intensive co-operation and long-term relation management. It was there that the neo-liberal interpreters inserted competition and quasi-market mechanisms, squandering their opportunities for a more integrative rationalization.

Effect-Oriented Environmental Administration in the Canton of Lucerne
In the canton of Lucerne, the Cantonal Office for Environmental Protection is one of the pilot organizations for New Public Management. To introduce these new management techniques, it was necessary to define the main products and services, quality standards, and main objectives. To do so, the Cantonal Office for Environmental Protection arranged a number of workshops with their traditional 'clients', target organizations, and partners in which these definitions

were drafted. On that basis, the internal organization was reviewed. A global contract with the cantonal government was then drawn up. It formulated the strategic political goals and presented a budget for reaching those goals. This is a telling example of how co-operative networking can and should contribute to policy formulation and implementation within the framework of New Public Management. However, by defining products, services, quality standards, and main objectives, the boundaries of the action radius of the Cantonal Office for Environmental Protection were circumscribed as well. The initial dialogue was not continued in the sense of a co-operative process for reaching specific impacts. Still, this is one of the few positive examples of 'intended' open planning.

Source: WOV Bulletin 3, January 1996.

Furthermore, under the pressure of the current financial *malaise*, other aspects of the original ideas will only be partially implemented. They will be either swept under the carpet or condemned to no more than lip service. In this way, for example, the mobilization of the innovative and rationalization potential of the workers on the administrative 'shop floor' seems to turn into the mere passing on of top-down austerity decisions to workers at the base of the hierarchy. Co-operative working teams and partners in co-operative networks become competitors bitterly fighting for their survival. The common purpose is no longer quality and effectiveness on the administrative shop floor, with respect to solving pressing environmental problems or developing the organization in view of future problems. Instead, the objective seems to be the short-term attainment of substantial budget cuts. The more the New Public Management scene in Switzerland is dominated by financial experts, the stronger this alarming tendency seems to get. A good accountant or controller is not by definition a good environmental manager.[8] This tendency also shows up in the one-sided orientation to financial indicators and measures. And -as mentioned before -it is behind the introduction of (quasi)markets, although empirical research shows that quasi-markets generate the desired rise of efficiency only in exceptional cases (Le Grand, 1993). It is also typical that Kuno Schedler (1995), in his book on Effect-Oriented Administration, devotes a chapter to the effectiveness of competition. With regard to the main thrust of the New Public Management concept, namely internal *and* external consensus-oriented co-operation, there are only a few

---

[8] As if the situation was not bad enough, some prominent private firms in Switzerland (and elsewhere) belong at the top of the list of bad examples by mixing the Lean Production concept, aiming at careful organizational development, with concepts for radical Business-Re-Engineering (see Kamiske & Füermann, 1995), or by maintaining a one-sided concern with short-term shareholder value, which turns lean, flexible, and innovative organizations into anorexic and dried-up milk cows.

dispersed references. In the latest version of the Effect-Oriented Administration concept, several important factors are neglected. The main gaps are the definition of value chains integrating *all* stages from natural resources to disposal, the inclusion of all relevant internal *and external* partners, and co-operation in strategic alliances. These are precisely the factors that would round out this concept in view of the hitherto experiences with environmental policy. Thus, the innovative and successful 'teeth' of the original concepts are being pulled out one by one, while the ideological thrust in the revised form is unmasked. The concept of Effect-Oriented Administration is on the way to becoming a one-sided concept for deregulation, privatization and short-term financial rehabilitation. In short: it is an example of the 'low road' rationalization, as described earlier in this chapter. In this way, we are wasting one of the most effective and promising instruments of modern environmental policy. Its loss poses one of the greatest dangers to a successful reorganization of the desired administrative reforms in general. It would certainly not be the first reform to be buried before it was completed. This danger seems very real. A cursory review of current trends might seem to refute these tendencies. But anybody who attentively and critically evaluates these tendencies can hardly deny their existence. This does not mean that there are no positive examples of successful administrative reform policies which implement the network and negotiated solution strategy in the framework of environmental policies. A random selection of success stories have already been described.

3.8 Summary

Let me summarise the main line of my argument. First of all, I have tried to describe the historical development of environmental policy implementation starting with the command-and-control paradigm and ending (for the time being) with the consensual steering model. To be able to adapt to these changing policy requirements, environmental agencies are undergoing a fundamental restructuring. In Switzerland, restructuring is taking place under the slogan of New Public Management or Effect-Oriented Administration. By returning to the roots of these concepts, I have uncovered the main reasons why these concepts represent a fundamental break with the past and have revealed some of their basic objectives. The main differences and advantages of these new concepts can be explained by adopting a more comprehensive conceptualization of rationality. Within the conceptual framework of New Public Management, it can be shown that discursive processes, -or, in the terms of this book, 'negotiated solution' strategies or private-public co-operative relationships- are fundamental to the success of environmental policy implementation. As an unintended consequence of this exercise, I have also put the negotiated solution strategy into a larger theoretical framework with respect to the general mechanisms for inter-personal and inter-institutional co-ordination of action. This should provide important clues for future research on these issues. Finally, I drew attention to the fact that exactly these basic elements and success factors of negotiation and discourse are in jeopardy. They are likely to be dropped because of the way these concepts are currently being implemented in Switzerland.

4. ENVIRONMENTAL PROBLEMS, ECOLOGICAL SCALES AND SOCIAL DELIBERATION

YRJÖ HAILA

4.1 Introduction

My purpose in this essay is to investigate the mutual relationships of some important conclusions on the nature-humanity issues that have been brewing, mostly independently of each other, in recent years. These conclusions stem from political and sociological studies as well as from an increasingly nuanced view of environmental science. I connect these developments to the perspective of developing social deliberation in environmental policy and politics.

The environmental 'awakening' of the late 1960s and early 1970s defined the environment as a political issue. However, it was less clear what kind of politics this precisely implied. Early programmatic declarations covered the whole spectrum of political philosophies from authoritarianism to radical egalitarianism; useful overviews are given by Dryzek (1997) and Paehlke (1989). Much has happened in political thinking since the 1970s. Although a wide range of views is still upheld in environmentalist movements, authoritarian prescriptions have, by and large, lost their credibility. A consensus has been growing on the importance of democracy for efficient addressing of environmental problems. (Doherty & de Geus, 1996; Lafferty & Meadowcroft, 1996; Mathews, 1995).[1]

However, the political implications of the environmental awakening reach further. Environmental problems stem from dynamic processes in which society and nature are inseparably mixed together. Consequently, the environment is not just an 'external fact' to be considered among other 'external facts' in decision-making. In contrast, the dynamic interpenetration of society and nature presents us with ever-changing complexity, spatial and temporal variation, and unanticipated consequences from apparently simple interactions which environmental decision-making procedures should accommodate. This idea has

---

[1] The ideal of democracy is not shared by all environmentalist groups, for instance, by right-wing radicals in Germany and authoritarian and centralized organizations such as 'Earth First!' in North America (Lewis, 1992).

P. Glasbergen (ed.), Co-operative Environmental Governance, 65–87.

been dubbed 'ecological rationality' (Bartlett, 1986; Dryzek, 1983, 1987). The need is not only for new administrative 'tools' to be used in an instrumental fashion but for new kinds of procedures; the practical experience is reviewed in Glasbergen (1996). The idea that decision-making should accommodate dynamic change implies that the divide between policy and politics becomes blurred. Apparently minor management decisions may have large political consequences; the other way round, the success or failure of specific management procedures may depend on political dynamics in society at large.

Another important development has taken place in the sociological understanding of environmental problems. 'Constructionism' -that is, the view that environmental problems are not read directly from nature but are 'constructed' through a social process of recognition, definition and evaluation- has come to rule the day in the 1990s. The arguments are summarized in, for instance, Hannigan (1995), Milton (1996) and Yearley (1992). Some degree of constructionism is supported by common-sense arguments. A problem is a social category; hence, problems are necessarily recognized within culture and articulated and communicated in language. What could a 'serious problem' that nobody knows about possibly be (Haila & Levins, 1992; p. 226)? The constructivist assumptions on which different views are built vary, but this is not important for the overall picture.[2]

In the following, I accept what N. Kathrine Hayles (Hayles, 1995) has called constrained 'constructivism'. By this she means that humans similar to other sentient beings construct their own world from their own perspective, but this construction is conditioned by objective constraints. However, we cannot definitely know the constraints. We only experience them in our activity, usually indirectly and *post factum*. This perspective gets support from case studies on how particular environmental problems have arisen into general consciousness (Taylor & Buttel, 1992; Wynne, 1994).

Constrained constructivism implies that not only finding solutions to environmental problems but also recognizing them in the first place is socially mediated. But this means that environmental policy cannot start from an assumption of unambiguous problem definitions either. This brings constructionism into close resonance with political thinking as, indeed, is shown by the work of Maarten Hajer on the role of discursive formations in acid rain controversies in England and Holland (Hajer, 1995).

Such a connection is believable on purely political grounds, too: the way of defining problems has implications as to how to start solving them (Haila & Levins, 1992). Furthermore, the ability to define issues is a critical determinant

---

[2] The critical distinction here is between the construction of ideas vs. objects, as Ian Hacking (1997) reminds us in a sobering comment on the discussion.

of social power. In the early stages of environmental awakening, for instance, pollution brought into focus the question, *Whose* health matters? Global problems such as climate change seem more equitable in their effects than local pollution. But this is a misperception: the issue of environmental justice is by no means off the agenda, particularly in the developing countries. Political institutions and socio-cultural processes are in a dynamic interaction when environmental problems become identified and evaluated. In other words, problem construction is constrained by social and political factors in addition to physical ones. Furthermore, both types of constraints are equally 'natural', as I will argue below.

4.2 The role of science

The conclusions presented above mean that the recognition of environmental problems and the shaping of policies to deal with them take place in a closer and more dynamic interaction than most of us would have accepted three decades ago. As a consequence, the role of science needs rethinking. An image of science built upon a strict subject object dualism is outdated, however firm the foundations on which this image is built. The environment simply cannot be 'externalized' outside of the social sphere in such a way that unambiguous standards for human affairs could be derived by research (Haila & Levins, 1992; Dyke, 1997). Evaluation of the environment is a social process through and through.

Moreover, scientists themselves are participants among other participants in the process of evaluation. They often promote, consciously or by default, their own role through sociological mechanisms; Yearley (1989) has analysed this in detail. This naturally also affects the conclusions scientists draw from their research, which makes controversies around environmental science unavoidable. No conscious distortion from the part of scientists themselves is needed. As a type of social practice, science is predetermined by a whole variety of social and philosophical conditions (for ecology, see Haila & Levins, 1992; Levins, 1996).

Processes determining the quality of the environment are 'mixed', socio-ecological, nature-cultural, 'hybrids'.[3] Caldwell (1970, p. 20) came to a similar conclusion in the early stages of environmental awakening through perceiving in environmental literature "a growing tendency to see man/environment relationship as a policy framework within which many specific problems can best be solved." The question is, How to conceptualize 'man/environment relationship' in a fruitful way? An 'objectivist' idea of science

---

[3] Bruno Latour (1993) brought the idea of 'hybrids'to attention in science studies, but the idea has been germinating for some time in the works of several authors; see, particularly, Haraway (1991).

does not stand up to the challenge. A formula of environmental science finding out 'the facts' that politicians should take into account simply is not adequate. Instead, the dualism inherent in this very formula needs to be discarded. Traditional ecological research is weak on this issue because of its commitment to an objectivistic model of knowledge.[4]

Several paths have been opened in recent years on modeling socio-ecological complexity. I will refer to some of this literature below. One of the major problems in this endeavour is to identify dynamic boundaries which separate different, relatively autonomous subsystems of the 'man/environment relationship' from each other. The environmentalist slogan "everything is affected by everything else" is not literally true, because different processes are dynamically independent of each other to variable degrees. However, criteria to use in an analytic decomposition of the 'man/environment relationship' are difficult to come by. One option is to proceed 'phenomenologically', by investigating variation in patterns of socio-ecological processes in space and time. This is what social historians, particularly of the French Annales School, have been doing for some time. Below, I will introduce the notion of an 'ecosocial complex' and use it as a conceptual tool for separating analytically relevant patterns from each other. There are limits to a phenomenological approach because the society/environment system is descriptively complex (Wimsatt, 1974).[5] Nevertheless, pattern descriptions, if successful, will help to identify recurring features of socio-ecological dynamics and to make enlightened guesses on how the dynamics work.

These considerations, then, introduce the issues I connect in this essay. A unifying theme is the potential of social deliberation in improving environmental governance. I argue that the potential germinates from all the strands of thinking I have reviewed above. To develop the argument, I begin in the next section by evaluating the legacy of the 'environmental awakening' of the 1960s/70s. In the two following sections, I introduce issues of socio-ecological dynamics and the role of science. In the last section, I address aspects of social deliberation, including some open questions.

---

[4] A typical form of the dualism in ecological basic research is the unquestioned view that only 'untouched nature' is worth serious consideration (Haila, 1992).

[5] Wimsatt's example of descriptive complexity is a developing insect, the point being that if the insect is imagined divided into parts using different criteria such as cell type, tissue composition, physiology, developmental fields, and so on, the resulting 'decompositions' look very different from each other. Decompositions of nature – society complexes using different criteria would lead to a similar conclusion.

4.3 Unity or differentiation? On the dynamics of the 'ecological crisis'

The environmental awakening of the 1960s/70s was, with hindsight, amazingly quick and internationally synchronized. An important role was played by the media by bringing into public consciousness pollution cases which obtained symbolic significance as local 'eco-catastrophes'. Important books such as Rachel Carson's Silent Spring (1962) were quickly translated into other languages. The environment also emerged as an issue of public policy. In a 'predecessor' country such as the US the federal government began to pay attention to the environment in the mid-60s, and the National Environmental Policy Act (NEPA) was passed by Congress in 1969. A 'latecomer' country such as Finland received a decisive push to establish governmental bodies for environmental protection strictly from the outside, through the European Conservation Year of 1970 declared by the European Council. A Ministry of the Environment was finally established after one-and-a-half decades of political wrestling in 1983.[6] An environmental impact assessment (EIA) procedure was included in NEPA in the US, whereas in Finland EIA was included in the environmental legislation only in 1994. The dates of comparable decisions in most other industrial countries are somewhere in between.[7]

The suddenness of the 'awakening' does not quite accord with the fact that many kinds of environmental problems, including some similar to those that were instrumental in triggering the awakening itself, had been known decades, even centuries earlier. Some aspects were novel, particularly the chemicalization of production and pesticide problems, but one can hardly identify a straightforward cause of the awakening in the material appearance of the problems. This basic constructivist point is, I think, well taken (Hannigan, 1995).

This gives rise to the following question: What actually was *new*? The idea that culture and nature are somehow in interaction was definitely not new (Glacken, 1967). Even the idea that human culture acts on the earth as a geological force was clearly formulated by George Perkins Marsh, in 1864, in his 'Man and Nature, Or, Physical Geography as Modified by Human Action' (Marsh, 1965). Geographers started long ago pursuing this point further; a particularly important

---

[6] Storsved (1993) gives an account of the Finnish case, but she attributes too much causal power to environmental ideologies.

[7] This variation among countries is highly interesting in its own right. I would venture to say that it reflects differences in national political cultures on the one hand, and in traditions of resource management on the other – the latter factor varying across countries as a function of what kind of resources there are, how important they have been and what political forces have been in change – but adequate comparative analyses seem to be missing. A timelag of 15-25 years in the early stages of environmental policy across countries will, of course, appear insignificant later on, but the political and cultural inertia probably remains.

compendium of their work was Thomas (1956).

I think a crucial feature in the ideology of the current 'awakening' is unification of most heterogeneous problems under the general heading 'the environment'. Various problems came to be identified and interpreted within a unified framework. As an example, consider the following characterization of the spectrum by Carolyn Merchant (1992, p. 17): "From Chernobyl radiation to the Gulf War oil spills; from tropical rainforest destruction to polar ozone holes; from alar in apples to toxics in water, the earth and all its life are in trouble".

The perception that all sorts of environmental problems are connected in a unified environmental (or ecological) crisis is built upon a metaphysical belief that the human environment is a unified whole. As pointed out by Ingold (1993), such metaphysics are in good agreement with the objectivism of traditional scientific thinking. In modern environmentalism, the human/environment system is viewed, as it were, from a distance, and the system is analysed as a unified whole. A global view of environmental problems is a view of the globe from a spaceship.[8] In addition, there have been social and political dimensions to the crisis perception. It has been reinforced by the dimension of political ecology, i.e., the effect of the environment on the real lives of real people. Natural hazards and socially induced hazards merge increasingly together, and the result is an increasing vulnerability of people to environmental changes and threats. Additional reinforcement has come from the unfolding of what Ulrich Beck (1986) has called 'risk society'. Beck argues that the rise of environmental issues connects with other changes in the modern industrialized society and that these together bring about a new type of social insecurity. Ulrich Beck's concept of 'risk society' is multi-dimensional and metaphoric. As regards environmental concerns, it is symptomatic that the very symbol of 'risk society', Chernobyl, was actually contingent; the text of Beck's book was finished before the accident. The concept has also changed in ways that need not concern us in this context. It seems to me, however, that a basic feature in the cultural archeology of the 'environmental awakening' is *increase in uncertainty*. This is captured by Beck's idea -and expressed with utmost clarity, for instance, in the introductory chapter to his original book.

A logical consequence of increase in uncertainty is that uncertainty gradually appears as the only certainty. This logical link establishes a close connection between the ecological crisis and crises of individual life in modern society (Beck & Beck-Gernsheim, 1994). In a 'risk society', there is a totalizing thrust to risk

---

[8] Powerful metaphoric expressions were used in article and book titles such as "The economics of the coming spaceship earth" by Kenneth Boulding, or "Only One Earth" by René Dubois and Barbara Ward. Caldwell (1970:3) started from a more mudane case and recommended that his readers see "the congested freeway as a problem of man/environment relationships".

consciousness, through the following dynamics: -Question: How to go about trying to solve an unsolvable problem? Answer: Construct the problem in such a way that it appears solvable (Haila, 1997; Haila & Heininen, 1995). Order is the opposite of chaos; it is a prerequisite for governability, controllability, tractability. Thus, a perception of order in both nature and in society is a prerequisite for coping with the environmental crisis at all.

The merging together of the idea of an all-encompassing ecological crisis and the question of social order gives postive feedback to the totalization of the 'environmental issue'. This is harmful from both sides, as it were. That is, not only are various processes of nature dynamically more heterogeneous and relatively autonomous than is ordinarily thought, but this is true of social processes as well. The very notion of a 'society' is another historical contruct (Ingold, 1996). In the next section I turn to the question: How can socio-ecological processes be analyzed?

4.4 Socio-ecological dynamics

Scaling in ecology
Ecology is a science of interactions. In 20th-century educated minds, this implies 'systems'. However, there are many alternative ways to conceptualize a system. A critical divide is whether interactions are assumed as linear or non-linear.[9] Linear (Newtonian) systems are scale-invariant; one can move from any level to any other without change in the structure of the interactions which make up the system. In non-linear systems, this is not true (Rosen, 1989). Processes have their characteristic, restricted spatial and temporal domains. Dynamic relationships become complicated. Cause effect chains across dynamically separated domains cannot be deduced by the Newtonian linear model we are accustomed to use in idealized mechanical systems.

The realization that scaling is a critical issue in ecology is fairly recent. It has become clear that space and time in ecology are not *a priori*, uniform Newtonian concepts. In contrast, they are constituted by the multitude of ecological processes themselves (Haila & Levins, 1992). The notion of 'physiological time' is an example of temporal scaling: the maturation of pine seeds in boreal latitudes depends on temperature, and the time required can be assessed as degree days, a cumulative sum of days that are warm enough, but not directly by calendar. The notion of 'home range' is an example of spatial scaling: the home range of a pack

---

[9] This is an oversimplification but sufficiently accurate for the purpose of this chapter. The 'chaotic' behaviour shown by many types of non-linear systems is subject of a vast literature; clear expositions of the issue of linearity vs. non-linearity include Kellert (1993) and Nicolis & Prigogine (1989).

of wolves is more adequately measured in terms of resource distribution than by using standard area units.

The interest in scaling has coincided with an interest in what has come to be called 'hierarchy theory' (Grene, 1987; Salthe, 1985). That life is organized in a hierarchical fashion is an ancient and phenomenologically straightforward idea. An ecological hierarchy, for instance, might include organisms, populations, communities, ecosystems, biogeographic zones (biomes) and the whole biosphere as organizational levels. Hierarchy theory deals with general patterns of dynamic relationships between different organizational levels. The connection between scales and hierarchies is evident: entities on different levels of a hierarchy scale differently in space and in time.

There are good reasons to think that such complexity, which implies a distinction between different scales, is an ontological feature of the world (Salthe, 1985; Wimsatt, 1994).[10] However, there may be different kinds of hierarchies, maintained by different kinds of processes that do not necessarily coincide. In ecology, O'Neill et al. (1986) drew a useful distinction between two aspects of ecological scaling, namely, structural and functional. The former refers to 'entities' such as organisms and populations, the latter to 'processes' such as flow of energy and circulation of matter. This distinction is not clear-cut, but it is useful, nevertheless, in that it suggests that there may be no one-to-one mapping between ecological 'structure' and 'function'. For instance, distributions of organisms and populations do not directly map to patterns of energy flow and nutrient cycling.

Inter-level analysis poses particular problems in systems that are non-linear and hierarchically organized. Stan Salthe (1985) has elaborated useful concepts for this purpose by drawing a distinction between a 'focal level' on the one hand, and 'initiating conditions' one level down and 'boundary conditions' one level up on the other hand. This is his 'triadic scheme'. Entities on a particular focal level are continuously maintained and reproduced by initiating conditions on the level one step down, and kept within bounds by boundary conditions on the level one step up. For instance, one might take a human organism as an entity on the focal level and figure out what are the necessary initiating conditions (largely inside the body) and boundary conditions (largely outside the body). The choice of a

---

[10] Does this sound like ontology talk? I am not worried, for two reasons. First, any description of the world implies an ontology (as Quine, 1953, among others, has pointed out); that is, an ontology for the descriptor (us), in distinction from a Kantian ontology 'an sich'. In this spirit, scales arise because every descriptor (including us) uses an 'observation window' in their various interactions with the world (Rosen, 1977). Second, it is particularly important to get rid of established ontologies such as the Newtonian time-space uniformity – which was adopted by Kant among his a priori truths. This being the case, it is particularly important to build deliberate vocabularies that imply alternative ontologies. This, of course, is what Hegel did to Kant.

particular focal level depends on the purpose of the analysis.

This differentiation allows useful insight into how change may occur in biological entities on a particular focal level. For instance, the state of a human organism may change as initiating conditions change (e.g., due to poisoning or disease), or as boundary conditions change (e.g., due to variation in temperature). As suggested by Salthe (1985), these are asymmetric: rapid and dramatic changes in ecological entities such as populations and communities are usually triggered by changes in boundary conditions such as climatic variation, habitat destruction, soil acidification, etc. However, there is always interaction between levels. Initiating conditions and boundary conditions are 'internalized' in the same system of interactions that allows the maintenance and reproduction of entities on a particular focal level. An example of such 'internalization' is the physiological adaptation of organisms, including humans, to geophysical conditions on the earth. Any adequately specified triad can be analysed as a dynamically interdependent, albeit not linear, whole.

Constructing triads may help to get the scales right in analysing human-induced environmental change. Take the eutrophication of a lake, a relatively simple case. Initially, nutrientflow into the lake from human sources (called external load, a change in boundary conditions) leads to an increase in nutrient level in the water body, but this is immobilized into a sink in the bottom sediment. After some time has lapsed, exceptional weather conditions (another change in boundary conditions) may get these nutrients to dissolve from the sediments (this is called internal load, a change in initiating conditions). This induces all sorts of change in plankton growth (another change in initiating conditions), and eventually an algal bloom results. It is quite likely that when nutrients accumulate in the bottom, this chain of events is repeated more and more frequently. Nutrients stored in the bottom sediments as well as changes in the structure of the fish community of the lake become new boundary conditions for the dynamics of planktonic growth in the lake.

Lake eutrofication is an exemplary case of ecological change induced by such human influence that has cumulative effects. In every particular case of eutrofication, both spatial and temporal scales are specified. Phenomenologically, they are the spatial extent and temporal duration of the changes that occur. Behind these phenomenological scales are corresponding processes, ordered on different levels of organization, viz., microbial reproductive loops in the lake nanoplankton on the lowest, and water circulation in the drainage system in which the lake is a part on the highest level. In the process of eutrofication, the dynamics of human-induced effects and ecological processes that would occur in the lake in any case merge.

This leads to a rule of thumb: the correspondence of scales determines what kind of human-induced change is of consequence in a given ecological system.

The rule helps to solve some apparent paradoxes concerning ecological relevance of human-induced change: an intensive change on a limited scale is (probably) insignificant, except for the individual organisms subjected to it, whereas a hardly observable change occurring uniformly on a large scale is (potentially) disastrous, although the consequences might not be visible in the lifetime of individual organisms (Haila & Levins, 1992, pp.182-90).

Ecosocial complex: introducing the concept
I have argued above that a basic feature of human-induced change in nature is the merging of human-induced and natural dynamics on a shared scale. This principle is straightforward on small scales. The chainsaw of a lumberjack and the bullet of a hunter disrupt physiological processes of the pinetree and the elk on the level of an individual, but it requires a lot of chainsawing and shooting to disrupt population dynamics of trees and mammals, let alone ecosystem processes in which trees and mammals take part. No single human being can through their activities achieve relevant consequences on such scales. Some notion of social agency is needed. The question is: What concepts to use to analyse the merging of dynamics on larger socio-ecological scales?

Fernand Braudel (1984) drew a distinction between three historical time scales, namely, 'geographical time', 'conjuncture' and 'individual time'. These temporal scales imply corresponding spatial scales. Geographical time is expressed as long-term secular change which is flowing in an almost uniform fashion over continent wide areas with civilizations as units. Conjunctures fluctuate 'within' geographical time, on spatial scales that vary according to the nature of each conjuncture. Individual time is constituted by life-cycles of single human beings and their immediate social relationships. Individual time is, of course, constrained by conjuncture, and in an undetectable fashion by geographical time flowing in the background undetectable because hardly any effects of geographical time can be experienced during the life-span of individuals.[11]

Haila and Levins (1992) introduced the concepts of 'ecohistorical period' and socio-ecological 'formation' to describe socio-ecological change in a long historical perspective. Such change is, of course, real and largely irreversible as has been known at least since Plato's Critias (Glacken, 1967, p. 121). However, long-term secular change is not uniform, although it may so appear.[12] Ideological constructs reifying long-term change such as 'Raubwirtschaft' or a Malthusian view of human population growth are examples of such misperceptions. Dynamic change occurs on the time-scale of conjuncture and is often triggered by events

---

[11] This offers an intriguing analogy with patterns of change in ecological systems (see Haila & Levins, 1992; Haila, 1995).
[12] This offers another important analogy with uniformitarian thinking in geology and biology.

occurring in individual time. For historical descriptions of such multilayered processes of historical change, see, e.g., Cronon (1985), Taylor & García-Barrios (1997) and Wolf (1982).

Conjunctures bring about locally specified ruptures in long-term, secular trends. In the following, I introduce 'ecosocial complex' as a conceptual tool for describing ecosocial processes on the temporal and spatial scale of conjuncture (see also Haila, *manuscript*). The aim is to identify ruptures in a descriptive, narrative or 'phenomenologica' mode. Detailed descriptions of patterns of change in ecosocial processes help to tease different, dynamically important elements apart. This task of making detailed descriptions also forces us to draw important analytic distinctions and define criteria for the identification of ecosocial change in the first place. That human-induced and natural change are not readily distinguishable from each other should be everyday knowledge in the era of climate change.

The term 'ecosocial complex' refers to any system of material flow from a natural source to a sink taking place under the influence of human social activity. That ecosocial complexes are 'socially mediated' means that the social activity of human agents dominates the dynamics of change in those complexes through some particular variable(s). For instance, in forestry -which I will use as an example later on- removal of timber is a basic characteristic in the dynamics of managed forests and it is, of course, completely human-induced. What sort of timber has been removed, and how this has actually happened, has varied depending on demand, economic factors, management schemes, technology and cutting practices. In addition, there has been variation in the social organization of timber removal: who did the job, who provided machinery and tools and gave instructions, where did knowledge come from, who got the profit, etc. Such social configurations have an integral role in the dynamics of the whole complex; they are not only background factors influencing the complex from the 'outside'.

I prefer the term 'complex' over 'system' to avoid Newtonian connotations. An ecosocial complex must be reasonably bounded to be identifiable, but the degree of boundedness is relative and ultimately a practical question. What matters is that considering processes within a particular ecosocial complex separate from the processes in its surroundings increases our understanding of the dynamics of the complex. It is particularly important that we be able to identify dominant dynamics, the characteristic scale in which these occur, and whether dynamic changes arise from social or natural sources.[13]

---

[13] In geophysical systems, this identification is sometimes straightforward. For instance, the very idea that radiative force brought about by greenhouse gases changes the earth's climate is another way of saying that radiation balance determines the dominant dynamics in climatic change. A well-known problem here, though, is that we do not know how predictable the changes are in regional

Bounding is relatively straightforward in the case of spatially isolated units such as distant islands; Easter Island might serve as a paradigm (Ponting, 1991). In such conditions, natural sources quite obviously determine the ecosocial dynamics. If soil erosion takes over on the mountain slopes of an isolated island, nothing much can be done. However, the socio-ecology of such isolated localities cannot be generalized -influences and interactions across long distances have been a dominant theme in human history ever since the beginning (Braudel, 1982; Wolf, 1982). On a non-isolated island an obvious way to counter the effect of soil erosion is to increase trade.

Another set of criteria might be derived from ecological boundaries for instance, watershed basins are to a certain extent ecologically separate from one other, in mountain regions more so than in lowlands. While this gives good criteria in some cases -Braudel's (1972) analysis of the mountain hinterlands of the Mediterranean comes to mind- this is not a generalizable rule either, because different human activities cross such boundaries to variable extent.

Ultimately, it is questionable whether natural boundaries are particularly useful in identifying ecosocial complexes. Since the dawn of history, societies have crossed natural boundaries. The point of trade is to exchange locally available items for locally unavailable ones. This presents us, hence, with a need to come up with concepts that capture the social dynamics of socio-ecological processes. I think the concept of 'social practice' is a promising candidate. Social practices are important for sociocultural dynamics, identity and social agency (this builds upon MacIntyre, 1985; see also Grene, 1985).[14] Furthermore, social practices often merge with particular natural elements through the exploitation of particular resources, which results in a closely interlinked, mutually conditioned process.[15] In a word, social practices are at the core of ecosocial dynamics.

A distinction between 'first nature' and 'second nature' offers another useful conceptual device for socio-ecological analysis (Smith, 1984; Cronon, 1991; Dyke, 1988). The distinction stems from Cicero (Glacken, 1967, p. 145) and is conceptually staightforward: 'first nature' refers to entities and processes that

---

weather patterns messed up by oceans, mountains, cloud cover, etc. When a dynamic change occurs, it is a matter of perspective to identify the 'cause'. For instance, the Indonesion forest fires of the early autumn of 1997 can be attributed to El Nino – provided extensive use of fire in land clearance in Indonesia is regarded as 'normal'.

[14] MacIntyre (1985; p. 187) puts it as follows: "By practice I am going to mean any coherent and complex form of socially established cooperative human activity through which goods internal to that form of activity are realized in the course of trying to achieve those standards of excellence which are appropriate to, and partially definite of, that form of activity, with the result that human powers to achieve excellence, and human conceptions of the ends and goods involved, are systematically extended".

[15] The notion of 'affordance' which originally comes from J.J. Gibsons's ecological psychology offers potential for analysing this aspect; see Grene (1985) and Ingold (1992).

occur without human influence and 'second nature' to elements produced by previous human activity. Both 'first nature' and 'second nature' constrain current human activity with equal force; there is no qualitative difference in this regard. Barge transportation has to keep to waterways; whether these are canals or rivers is of secondary importance. Besides, natural and human-induced processes are mixed as far as the maintenance of waterways goes: both canals and rivers used for barge transportation require draining and regulation.

Dyke (1988) made the important point that 'second nature' does not only comprise physical entities and structures but also sociocultural factors such as institutions, rules, customs and habits. An established rule is a constraint on human activity or social practice similar to a physical barrier. As can be easily seen, these concepts bear an interesting relationship with Braudel's temporal scales. Institutions such as religions and administrative innovations unfold in geographical time; rules and customs may change on the time scale of conjuncture; and habits correspond to individual time.[16]

Ultimately, the decision on where to draw boundaries -i.e., which processes to include 'within' the complex and which to stabilize as boundary conditions- is contextual and depends on the purpose. For instance, as regards forestry, one might include cutting methods in the complex and 'freeze' the present structure of production as a boundary condition; this seems fair on the time scale of a few years. Or, alternatively, one might include both cutting methods and the structure of production in the complex and consider conjunctures on the world market a boundary condition; this seems fair on the time scale of a decade or two.

How does all this help? The concept of ecosocial complex can be used to draw distinctions among different factors in ecosocial dynamics. Hence, it can also help in identifying factors that influence and modify socio-ecological change. To give an example, I elaborate in the following briefly upon a case of human-induced environmental change that has undeniable ecological consequences and is based on a social practice but, on the other hand, lacks 'natural' boundaries altogether: forestry in Finland.

4.5 Ecosocial complex: an application to Finnish forestry

There is an abundant literature on the history of forest use in Northern European countries; Åström (1978) gives a good overview of the Finnish case. Haila and Levins (1992) described the broad outline. I will therefore focus on changes at the

---

[16] The distinctions are certainly blurred but terminological details need not concern us here. The important point is that this set of concepts has potential for analytic purposes; see Dyke (1988) and Murphy (1994).

level of conjuncture during the period of industrial forest use.[17]

The beginning of mechanized forest industries, driven by steam instead of water power, was sudden. It was triggered by changes in regulatory legislation in 1857 in what at that stage was the Grand Duchy of Finland, an autonomous part of Imperial Russia. The early growth of these industries followed international conjunctures. Major industrial companies in both sawmill and paper and pulp industries were established in the 1860s and 1870s, partially by foreign (mainly Norwegian) capital. The branch was highly monopolized from the very beginning. Company founders acquired large forest properties on advantageous terms, as the economic value of forested land was not yet known. Furthermore, the population in the Finnish countryside faced hardship in the aftermath of a serious misharvest and hunger catastrophe in the early 1860s. This, however, triggered public indignation; the trend was reversed through legislation after Finland got independence from Russia in 1917. A close link was retained between farming and forest ownership. In the sparsely populated eastern and northern parts of Finland, the state remained the most important forest owner. In these parts of the country, workers were needed in lumbering, particularly during the winter. A lumberman smallholder economy developed and remained intact until the 1950s (Rannikko, 1995).

An institutional framework for forest exploitation grew up during the second half of the 19th century. The National Board of Forestry was established in 1859; higher education started in 1862; and a professional organization, the Finnish Forestry Association, was established in 1877. Legislation was modernized to prohibit devastation of forests in the 1880s. The state started to subsidize and regulate forestry on private land through agricultural associations in the 1890s. Eventually this gave rise to a separate advisory organization for forest owners. The history up to post-WW2 years can roughly be divided into two periods (Donner-Amnell, 1991). First came the breakthrough of industrial forestry, up to the early decades of this century, driven by industrial interests and the slowly maturing public organizations as well as interest groups of professional foresters. This period was followed by a stabilization of sustained-yield forestry in the interwar years, driven by strengthening corporativism, which comprised the administration of the newly independent nation-state, industrial interests and organizations representing forest owners.[18]

A shift to intensive silviculture occurred relatively late, in the 1950s. This was driven by the corporativistic structures that took shape before the war. A major

---

[17] The sources I have used are largely in Finnish; the outline of the story is told in Donner-Amnell (1991), Haila et al. (1971), Lehtinen (1991) and Raumolin (1984, 1985).

[18] The sectoral corporativism around Finnish agriculture, and environmental management more generally, is characterized in Jokinen (1997) and Hukkinen (1995), respectively.

change occurred in cutting practices selective cutting was banned[19] and forest owners were required by law to use clear-cutting, soil preparation and sowing or planting in forest regeneration. Research on intensive silvicultural methods also replaced older research traditions which were more attuned to forest ecology at the Forest Research Institute. Intensive silviculture was supported by the government in remote parts of the country. A 'zero limit' beyond which forestry was not economically feasible was still known in Finnish Lapland in the mid-50s, but it vanished by the end of the decade.

Virtanen (1995) emphasized an important peculiarity of forest industries compared with other resource-extracting industrial branches: forest industries brought a money economy to remote, forested parts of the Finnish countryside (see also Rannikko, 1995). Cutting and transportation required many workers, and they were, of course, paid in cash. Forest owners earned money by selling timber. The growth and spread of forest industries occurred in close interaction with a dramatic social change in the Finnish countryside.

In the era of cost-effective silviculture, this influence has diminished, however (Lehtinen, 1995). Employment in forestry declined dramatically in the period after WW2 due to rationalization. The chainsaw was introduced in the 1950s, the transport of timber has shifted from rivers to roads since the 1960s, and mechanized harvesting has taken over the work of manual labour since the late 1970s. Operative planning in forestry is increasingly centralized in corporations. The operations are monitored almost in a real-time mode and undertaken by independent enterpreneurs. There is also a trend toward deregulation. It was partially triggered by environmentalist opposition against the forest legislation of the post-WW2 era, which set strict limits on what individual forest owners were allowed and not allowed to do. At the same time, the structure of forest ownership has changed: due to both inheritance and forest land speculation, an increasing proportion of forest owners live in the cities. This has two important consequences: incomes flow into the cities, and forest owners lose a personal relationship with their property. This presumably, in the long run, is conducive to following corporate-driven advice on how to use their forests.

Nature conservation has emerged as a relevant issue in forestry during the 20th century. Foresters were active in the early stages of nature conservation around the turn of the century (Lehtinen, 1991), similar to albeit later than in Germany (Radkau, 1997). Modern environmentalism reached Finland in the late 1960s, and

---

[19] Selective cutting had been practiced particularly on private lands all over southern and western Finland for decades in the shape of removing at each cutting only high-quality timber; there is good evidence that this had caused stand deterioration over vast forest areas. A case can be made, however, that a different, ecologically grounded system of selective cutting would avoid the shortcomings of the traditional type (but note that most managers would disagree with this point).

the forest question has been at the center of attention, for obvious reasons (Lehtinen, 1991, 1995). The first wave of public, environmentalist criticism of intensive management arose in the early 1970s, and a second wave, which included forest activism, in the mid-1980s (the latter one is the focus in Lehtinen, 1991). In the 1990s, the situation changed again, as concern about forest conservation -largely under the heading of biodiversity protection- became institutionalized. This issue was taken up first by forest corporations, then by state administration, as a result of increasing pressure from Central European markets (Lehtinen, 1995). New management guidelines have been given by all important organizations dealing with forestry as well as major companies, but their efficiency remains to be seen.

This brief historical outline allows two kinds of conclusions. First, there have been ruptures in ways of forest use in Finland, triggered mainly by international conjunctures but driven by national actors. These ruptures have changed the face of Finnish forests. This is because, by and large, in this context the whole country can be regarded as a single unit. Consequently, changes in management methods at each stage have swept over the whole country in a fairly uniform fashion. Changes occurring consistently over large land areas will bring about ecological consequences (Haila & Levins, 1992, p. 190). On these grounds, one can claim that the ecosocial complex of Finnish forestry has changed at the junctures, although it has always taken some time before the consequences have become visible.

The present situation is multilayered: efficient, mainly state-driven silviculture introduced in the 1950s continues as the dominant current, but new incentives mediated through the market as well as conservationist pressure may bring about changes in management practices. How far-reaching such changes will be remains to be seen. One likely scenario is that the corporate sector, backed by the state administration, will increasingly come into control of the most profitable forestry operations, under the slogan of 'international competitiveness', while some part of the forested land will be allocated for conservation and, possibly, local use. It is worth noting once more that an 'internationally competitive' forest sector tends to become uncoupled from local economic and social ties.

Second, the relative weight of different social actors has varied from stage to stage. The early phase of industrial forest use was dominated by the interests of the company founders, but quite soon also forestry professionals and state officials appeared on the scene. The close link between forestry and farming was resumed through legislation after WW1 and Finnish independence. Since then, agricultural and forest owners; associations have been deeply involved in forest management on a corporativistic basis. The strengthening of this 'alliance' required legislation which imposed strong restrictions on forest use by individual forest owners; this legislation particularly affected use during the implementation of intensive silviculture. Finally, conservationists have been on stage since the 1960s. Workers have never had much of a say in the shaping of forest management practices. Presumably, this is because permanently employed workers have formed only a tiny proportion of the total work force employed in lumbering; labour unions have, accordingly, been weak. In addition, the relations between workers and technicians and forestry professionals have traditionally been extremely authoritarian. Oral histories of lumbering work have convincingly documented this aspect of the social organization of forest exploitation.

The notion of an ecosocial complex emphasizes that sociology and ecology of forestry belong together; they cannot be separated from each other. A forest's nature is, of course, immediately modified by actual forestry operations, but changes in these are driven by socio-economic factors. Major influences have been effective on the international scale. As the forest sector has been export-oriented from the very beginning, international factors have had a dominant influence on the history. Up to the 1990s, this was mediated only by industrialists and governmental agencies, but recently conservation has emerged as an international factor, mediated by NGOs (Lehtinen, 1995).

A more fine-scaled analysis, based on the socio-economic organization of the Finnish forest sector, would be possible too. This might be based, for instance, on differences in the organization of the labour process: independent farmers in the south and west, lumbermen-landholders in the east and north (Rannikko, 1995). Variations on this scale of resolution would raise important issues of how to identify stakeholders and understand their motivations and goals in disputes over forest use. The term 'local population' is often used in environmentalist discussions in a typological fashion as if it would refer to a predefined natural type, but this is totally inadequate in any specific case (for Finland, see Lehtinen, 1991, 1995).

4.6 Problem assessment and scaling

My main argument thus far has been that the understanding of environmental problems requires adequate scaling of socio-ecological change. Let us keep in mind that the goal is to maintain life-support systems of the earth in such a shape that human existence remains possible. 'Stability' or 'constancy' are misleading surrogates for this goal. Nature is aways changed when it is used, whether this is done by humans or by other organisms. The real concern is that ecological systems retain their resilience, i.e., that the change remains within bounds.[20] For

---

[20] The term 'resilience' was introduced by Holling (1973); for resilience and ecological scales, see also Holling (1986).

us humans, a proper focus is on the consequences of our own activities. That there are different ecological scales to consider implies that ecological change needs to be evaluated against several alternative 'contrast spaces'.[21] To continue with forestry, an appropriate contrast space for assessing the consequences of a particular cutting operation is not 'no change vs. change'. More adequate alternatives would be, for instance, 'regeneration successful vs. no regeneration', or 'no harm vs. harm caused to forest organisms other than trees'. These alternatives imply different spatial scales. The former concern is, by and large, restricted to the site under consideration, whereas the latter means that forest composition on a larger regional scale needs to be considered as well (Haila, 1994).

However, all the criteria to be used in assessing ecological change cannot be read directly from nature. This follows from the intrinsic interconnectedness of culture and nature. I have argued earlier that social structures as well as cultural customs and habits act as constraints on nature use in a way analogous with ecological factors. This means that another set of contrast spaces for evaluating specific ways of nature use can be derived from social considerations such as, Do they support harmful vs. benign structures/ customs/ habits?

Grounds on which to evaluate and respect nature emerge in human consciousness. Different aspects of nature have different meanings in different contexts and for different human beings; 'values' considered in isolation from the cultural context are uninteresting abstractions (Brennan, 1992). This is in line with the constrained contructivism I discussed above. Established cultural values related to particular elements of nature stem partially from productive practices and can, thus, be connected with particular ecosocial complexes. The contrast between a native vs. a colonial perception of elements of nature, thoroughly documented from all parts of the world, bears evidence of such differentiation.

Ultimate criteria for assessing change in nature are not up to humans to invent. Instead, they stem from constraints that have an objective character. The task is to identify such constraints and give them a name -hence the importance of science. Note, however, that science is continuous with other types of knowledge. The notion of social practice is interesting in this connection. Knowledge of major constraints such as seasonality has become 'internalized' in human practices such as agriculture or fishing. Other sorts of constraints are not eqully visible. Furthermore, there may be ways to manipulate the constraints: irrigation changes the effect of seasonality on agriculture; the development of efficient gear changes its effect on fishing. But then social contraints on nature use tend to become internalized as well. For instance, land ownership has been a more important institution in Finland than in probably any other Western country and has a major

---

[21] See Dyke (1988) on the applicability of the term contrast space to scientific explanation.

effect on planning and implementation of conservation measures.[22]

Science is needed for addressing environmental problems, but what sort of science is useful? An implication of what I have argued above is that mere 'phenomenology' -recording the state of the environment in terms of pollution level or resource use- does not get us very far. Of course, this has been necessary as a first step in assessing the seriousness of environmental threats. Scientific expert councils on the environment were established in most countries early on, even prior to administrative bodies (Weale, 1992). Phenomenology -i.e., patterns of pollution and other types of environmental destruction- does not allow extrapolation outside the domain of observations, for instance, toward the future. The issue of the dying forest offers plenty of examples of alarm scenarios which turned out to be exaggerated.[23] More importantly than with phenomenology, systematic knowledge is needed of critical processes that human activities actually threaten to disrupt in nature.

The triadic model of Salthe (1985) I introduced above is useful in this context. One can identify the phenomenon of interest as a 'focal event' and then evaluate the event against the boundary conditions or initiating conditions of processes it affects. Let us return to the example I used earlier in this section, the cutting of a particular forest stand. A worry about regeneration can be formulated as the question, Does the cutting change the boundary conditions of forest growth? A worry about harmful effects on biota other than trees can be formulated as the question: Does the cutting change the initiating conditions of population or community dynamics on the regional scale?[24]

Obtaining processual knowledge about how human-induced disruption actually modifies natural processes presents methodological challenges. Rigorous experiments are seldom possible for the simple reason that we do not want to subject hypotheses about ecological destruction to experimental testing; we want to prevent it (Stewart-Oaten et al., 1986). Nevertheless, established scientific methods can be followed in impact assessment. Although a catastrophe such as the *Exxon Valdez* case cannot be repeated, impact assessment leans partially on knowledge of such particular ecological processes that can be studied experimentally (Wiens, 1996).

In terms of the social organization of research, obtaining processual knowledge implies that experience from past management operations should be used as basic

---

[22] This line of thought brings us back to the theme of 'hybrids' that I touched upon earlier.

[23] On the other hand, overstatements are understandable and unavoidable when attention is directed to a serious problem nobody seems to know or care about. Furthermore, a *post factum* claim that an environmental alarm was false is logically suspect in the same way as the statement that the "Peace movement was unnecessary because the war did not break out".

[24] There is plenty of evidence that both sorts of harmful effects often happen, but a more detailed consideration is not possible in this context.

data. This principle was called 'adaptive management' by Walters (1986). A growing literature attests to the need to develop an adaptive management of biological resources (Gunderson et al., 1995; Holling & Meffe, 1996). Furthermore, essential challenges in fundamental research such as the need to develop biodiversity monitoring schemes can be addressed through a similar approach (Haila & Margules, 1996). In a word, research should get closer to social practice. For this to succeed, people involved in practical management ought to participate in all stages of the research process, from defining the problems to evaluating and interpreting the results. This is a way to bridge the gap between generalizing knowledge of ecological processes and idiosyncratic experience of particular situations (Lee, 1993). This is also a way to substitute context-specific dynamic analyses for an amorphous worry about a unified environmental crisis. Furthermore, contextually specified and socially sensitive research can have a postive influence on research in traditional fields such as ecology (Taylor, 1997).

These conclusions imply that science cannot be an independent stakeholder in environmental affairs. Rather, different stakeholders back their claims by reference to whatever knowledge they think is relevant, and this includes variable amounts of strictly scientific arguments. Science cannot act as a neutral judge in disputes (Yearley, 1989). Fortunately, open disputes are not all there is to enviromental discussions. If some particular scientific findings are absolutely urgent, various parties may eventually join in and accept them as indicating a major constraint on human activity. This, after all, has happened in a relatively short time in the case of ozone depletion.

4.7 Deliberative environmental policy?

I have discussed the need for socio-ecological analysis primarily on theoretical terms, as a means of increasing our understanding of the dynamics of human-induced environmental change. However, the need can also be formulated in more practical terms: socio-ecological analysis helps to identify entry points from which social and cultural structures conditioning socio-ecological change can be influenced.

This brings environmental policy and politics into the picture. Socio-ecological analysis can turn practical when it helps to bound environmental problems in an adequate way. However, this is not a task for a master analyst; social actors should be brought in. The analyst might be able to help in framing interesting questions about problems that need to be solved. Problems are not natural entities, sitting out there and waiting to be detected. Problems are identified or constructed. The purpose of the first section of this chapter was to point out that this happens through a process which in itself is socio-ecological in character. The purpose of the second section was to point out that problem unification, under a perception of a general environmental crisis, is a pitfall that needs to be avoided. As a matter of fact, a case could be made that environmental issues as they appear in political disputes are inherently multifaceted and heterogeneous, due to the historically contingent nature of the process which triggers the politicization of specific issues.[25]

Environmental problems, once identified, can be evaluated further using the triadic formula of Salthe (1985). The boundary conditions and initiating conditions of such socio-ecological processes in which those particular problems are enmeshed, can be distinguished from each other and analysed separately. Boundary conditions are constituted by all the factors that predetermine ways of using nature, be they structural in a literal, institutional sense, or questions of ideology and world-view. Initiating conditions are constituted by factors that shape social agency, that is, specified social actors with their goals and motivations.[26]

Social deliberation would be valuable in this process in two ways. The first way is in problem identification. The goal is to end up with a common understanding of what kind of problems there are and what sort of measures can be used to address different types of problems. It is common knowledge that the perception of environmental problems has changed through the years. 'Old-fashioned' methods such as straightforward prohibition, retain their usefulness whenever values commonly agreed upon are directly threatened. This can be equated with such moral rules that have become ingrained in commonly shared ethical sensibility -for instance, the prohibition of child labour. In environmental policy, international agreements work partially in the same way, by presenting obligations that no nation-state should leave unfulfilled.

However, in most practical cases, the situation is more complex, and problem definition requires a different kind of deliberative consensus. John Dryzek (1995) has defended the view that democracy is communication and that nature should, and can, have a say in communicative processes. Of course, nature does not 'speak out' in a literal sense. Nevertheless, nature gives signals on what is allowed for humans and what is not. Every organism interprets and responds to signals that give this kind of contextual information on the environment. Humans

---

[25] This is analysed by Laine, Peltonen & Haila (manuscript) using the politicization of the environment in Tampere (Finland) as a case study.
[26] A note may be in order on the term 'initiating conditions'. The formulation draws a purposeful distinction with 'initial conditions', a notion used in mechanics. The point is that 'initiating' implies a continuous process of initiation, whereas 'initial' connotes an original push by, for instance, God's hand. Social agency is 'initiated' continuously anew.

are no different in this respect. The question is, how to systematize such interpretation.

My position is that nature's 'speech' can be brought forth in connection with human practices. Practices are constrained and shaped on a day-to-day basis by natural processes on which they depend. This is not straightforward adaptation but rather mutual interaction and coevolution, 'hybridization'. Artefacts, constrained simultaneously by nature and by culture, materialize in themselves some of this 'speech' (Grene, 1978; Dyke, 1994). In general, communication occurs naturally in activities closely bound to a particular resource base -for instance, agriculture, fishing and hunting. Literary works, for instance Moby Dick or the Sea-Wolf, bear evidence of this process. Not only nature but also social constraints are attested by these novels. A theoretical challenge is to understand how communication between human practices and nature might become 'audible' in modern, complex societies.

Another perspective emphasizing the role of deliberation is indicated by the question: Who are the participants? Deliberation should be open to everybody who is motivated to participate. This is a normative claim but can be substantiated, too: democratic practices tend to be reproduced and deepened by being used. The circle of participants can get wider through the constitutive role of democratic practices. By fostering a broader participatory basis, a greater variety of experiences are collected, and a greater variety of values are brought together.

The Finnish experience thus far primarily gives evidence of barriers to developing social deliberation.[27] A commonly shared explanation is that for historical reasons the position of civil servants has been exceptionally strong in Finland. The roots of this feature of Finnish society may go back to the 1610s and 1620s, when centralized state administration began to strengthen in Sweden. But the period when Finland was under Russian rule in the 19th century has had a major influence (Ylikangas, 1993). Hence, the strong position of centralized administrative bodies which have influenced, partially on a corporativist basis, resource use, road and waterway construction, and land-use planning. Traditions within environmental protection agencies are not insulated from such a burden of traditions either.

A final question is whether democratic deliberation can provide long-term strategic assets for addressing environmental problems. This question belongs to the realm of social philosophy. There are grounds for cautious optimism. Participation in environmental decision-making may encourage people to think of nature in new ways. Then, if nature can be given a voice in narratives on what it means to be a human being, this may further improve the motivation and

---

[27] Remember that EIA legislation was passed in Finland in 1994.

'capability' (Sen, 1993) of people to act. However, whether this perspective can turn into reality is an open question.

Part II

Experiences

5. THE DIVERSITY OF ENVIRONMENTAL AGREEMENTS

An international overview

BETTY GEBERS

5.1 Introduction

Environmental agreements have become a very popular and often discussed instrument of environmental policy. Surveys and studies that were performed recently showed that voluntary agreements are a vital element of environmental policy in most member states of the European Union.[1] But it is not only in Europe that agreements have become a popular form of policy-making - examples from Poland and the U.S. show some interesting approaches as well.[2]

A closer look at the agreements in several countries shows that voluntary agreements are not a uniform instrument. There are many varieties. Differences can be observed with regard to the policy context, the negotiation process, the participating parties, the contracting parties, the content of the agreements, the legal form and the monitoring activities. The activities that are usually subsumed under this term differ significantly. They range from self-commitments in Germany to more or less binding contracts in the Netherlands to 'clandestine' agreements in Belgium. The policy debate in different countries concerning this subject is thus not based on a uniform understanding.

This chapter illustrates the diversity of design and use of the instrument through an overview of the way it is applied in nine countries: Belgium, Denmark, France, Germany, the United Kingdom, Italy, the Netherlands, Poland, and the United States. It also identifies similarities that exist, in spite of all the differences.

Before that, the problem of formulating a uniform definition is discussed. Use of the same definition can greatly facilitate communication and research on the subject.

---

[1] See: European Environmental Agency, Environmental Agreements, vol.1, Copenhagen, 1997. Öko-Institut, New Instruments for Sustainability - The role of Voluntary Agreements - Interim Report, Darmstadt, Utrecht, London, 1997.

[2] See e.g.: United States Environmental Protection Agency, Office of Pollution Prevention and Toxics - A Progress Report: Reducing Toxic Risks through Voluntary Action, Washington, 1991.

P. Glasbergen (ed.), Co-operative Environmental Governance, 91–109.

5.2 Definition of environmental agreements

The question of how a voluntary agreement in the field of environmental policy can be defined has been addressed by several authors. In a 1997 study of environmental agreements by the European Environment Agency, the authors come to the conclusion that "there is no standard definition of environmental agreements, which are also known as voluntary agreements, negotiated agreements or convenants."[3] For the purpose of the study they decided to cover "those commitments undertaken by firms and sector associations which are the results of negotiations with public authorities and/or are explicitly recognised by the state." A study by the European Commission (DG III) on voluntary agreements has chosen the same scope, but notes that the term 'Voluntary Agreement' bears some inaccuracies. The authors suggest to use the term 'Environmental agreement' instead, but do not develop a definition of their own which matches the suggested terminology.[4] Neither study includes the goals and the environmental policy dimension of the instrument in its definition of the term 'Voluntary Agreement' or 'environmental agreements'.

In its Communication (Com(96) 561 fin.) the European Commission defines environmental agreements as

"a contract between individual companies and/or association of companies on the one hand and public authorities on the other hand, concluded with the aim to protect or restore the environment."

This definition is more precise, because it includes the environmental policy dimension. The definition reflects a certain preference for the legally binding contract[5] and it might exclude self-regulatory commitments and possibly also two-sided agreements that both sides would not refer to as a contract. In reality, these rather 'soft' agreements can be met more often than contracts. Therefore, a definition has to be preferred that also covers self-regulatory instruments provided that in the negotiating phase the state was one of the stakeholders.

---

[3] European Environment Agency, Environmental Agreements, vol.1,: Environmental Effectiveness, Copenhagen, 1997, p. 11.

[4] European Commission, Directorate General III.0, Industry, Study on Voluntary Agreements Concluded between Industry and Public Authorities in the Field of the Environment, Final Report - Annexes, January 1997, p. 9 (not published).

[5] See also Rehbinder, Eckard, Environmental Agreements: A New Instrument of Environmental Policy, San Domenico 1997, p. 14.

In its study "New Instruments for Sustainability - The Role of Voluntary Agreements" the research team first defined the core elements that characterise environmental agreements. These elements were synthesised into the following definition of an environmental agreement. It is:

"An agreement or an action of self-regulation which is voluntary in character, that involves stakeholders of which at least one is the state, that is either a substitute or that is a device for implementing or going beyond environmental law and policy, and that is aimed at sustainable development".

An agreement or an action of self-regulation

In some countries environmental agreements are drawn up as contracts, with both government/administration and the industry signing it. This would be the classic two-sided agreement. However, two-sided contracts are not signed in all countries. In Germany, industry associations commonly negotiate with the government, but the result is most often a self-commitment of industry. The government will usually offer some kind of relief in return for the commitment. Because the action of self-regulation involves government or administration it is in fact seen as an environmental agreement as well.

Voluntary in character

It is a specific element of environmental agreements that they are concluded voluntarily. This distinguishes the instrument from instruments where legal obligations exist. Very often a so-called 'voluntary' environmental agreement is signed because of a legislative threat behind it. Thus, it is not completely voluntary: the industrial party agrees, because it otherwise would have to bear a bigger or a less attractive burden. However, the parties have the choice to agree or not to agree; there is no legal obligation to join in. This distinguishes environmental agreements from other instruments like command and control laws or eco tax laws.

Different stakeholders

A self-commitment of industry to achieve a certain environmental goal cannot be called an agreement if it does not involve other parties at all (such as industrial code of conduct). It is a vital element of a voluntary environmental agreement that it involves stakeholders of different interests. However, the role of the stakeholders might vary. There can be intensive negotiations. One can also find agreements where the government/administration says that it will react with a certain counter performance.

State participation

If environmental agreements are indeed instruments of environmental policy, they will somehow have to involve the state. State participation is for this reason another vital element of an environmental agreement. Other parties might be involved as well. Especially at the local or regional level (Neighbourhood Agreements), municipal bodies and/or citizens groups might participate in the negotiation as a party.

Connection with environmental law and policy

As concerns the actual content of an agreement, a variety of objectives can be found in the examined countries.

Firstly, it can be a substitute for an environmental law (e.g., the government withdraws or modifies a legislative proposal).

Secondly, it can be a device for implementing environmental law and policy. (For instance, this would occur if the government has set a certain environmental target and asks industry to join in with a certain commitment.)

Thirdly, it can be a device for going beyond existing policy and legislation. (For example, this would happen when manufacturers agree to phase out a certain substance sooner than the Montreal Protocol calls for. It could also happen when a plant operator agrees to meet stricter environmental standards than provided by the law.)

Aim at sustainable development

An environmental agreement has to address the ecological dimension of sustainability. It should aim to contribute either to a reduced use of resources or a reduced release of pollutants.

National, regional, local level

Environmental agreements are recognized as an instrument of national environmental policy. This might be due to the greater public attention these agreements receive in comparison to regional or local agreements. An examination of the current practice in the different countries shows that the main elements of the instrument are similar on all levels. However, some differences between local and national agreements exist. The local agreements very often involve more negotiating parties. In addition, the design of the instrument seems to be more transparent on the local level. The enforcement of an agreement on the local level will often be less problematic and controversial, because its mandatory character might possibly be stronger.

5.3 International overview

In the following, an overview of the use of environmental agreements in nine countries is given. The overview summarizes the findings of detailed country reports which were conducted in the framework of an international research project by the Öko-Institut e.V., Darmstadt, in co-operation with the Foundation for International Environmental Law and Development (London), Stichting Natuur en Milieu (Utrecht), Centre d'étude du droit de l'environnement[6] and several other institutes.[7]

Belgium

The first agreements were adopted in the late eighties although contractual techniques already existed in labour law, pricing policy and consumer protection. An interesting feature in Belgium is that one of the country's regions, the Flemish region, is endowed with legislation dealing specifically with environmental agreements. The Flemish decree of June 15, 1994 provides a clear legal framework for environmental agreements concluded between the regional government and industrial organizations. It has been adopted to put a brake on the adoption of agreements, the content of which was unclear. In order to better control the move towards this new kind of instrument, the goal of the decree is to frame the future 'covenants' in a plain and transparent legal procedure which is binding for the parties.

The decree deals with:
- the accepted contracting parties;
- the extent of the limitation of the regulatory power of the government;
- the binding force of the agreement;
- the adoption process (the decree provides a communication for the regional parliament);
- the information process (publication in the official journal);
- the sanctions and control concerning the agreement.

No environmental agreement has been adopted under this decree so far.

Not more than ten environmental agreements, between public and private partners, are in force in Belgium today. They deal among other issues with:
- electricity-producing facilities (agreement involving both federal and regional authorities);

---

[6] Öko-Institut, New Instruments for Sustainability - The Role of Voluntary Agreements- Interim Report, Darmstadt, Utrecht, London 1997.
[7] The author would like to thank Jan Willem Biekart, Stefan Brendstrup, Ralf Jülich, Ruth Khalastchi, Delphine Misonne, Halina Ward, Jerzy Jendroska, Stefano Nespor and Sanford Lewis for their contributions to this section.

-   the glass industry (on air pollution);
-   the cement industry (on the reuse of waste);
-   the construction sector;
-   the detergent industry;
-   the pharmaceutical industry.

The enforceability of most of these agreements is weak and information about most of them (texts, results) is not easily available. A positive point, however, is that the instrument appears to create a dialogue between the parties and is a kind of platform for discussions on a regular basis.

Environmental agreements have, so far, not been put very high on the agenda of public authorities, due perhaps to various institutional hurdles (i.e., the sharing out of competence between the various authorities) and the uncertainties about their possible binding force (except for the Flemish region, see above). Nevertheless, the idea of 'Voluntary Agreements' has not been given up. The possibility to use this kind of instrument has been written into drafts of legislation to be adopted at the federal level and in the Brussels region.

Denmark
Different kinds of agreements have always been a part of Danish environmental legislation. To some degree this fact reflects the culture of compromise which is characteristic for environmental policy in Denmark. However, environmental agreements which are brought directly into use instead of by other possible means are a rather new phenomenon. The emergence of these agreements in Denmark is to be understood as the result of an intention to adjust environmental regulation in the direction of more efficiency. That is to be achieved through cutting down on bureaucracy and creating more flexibility and self-regulation. Such agreements may also be seen as a result of pressure from representatives of industry and trade, who wanted to reduce command and control regulations.

All Danish agreements have the character of negotiated agreements. Though the Ministry of the Environment and Energy is not always an official party to the agreement, the agreements nevertheless have been initiated by or directly negotiated with the authorities. Agreements have mainly been used in relation to rather limited problems and specific circumstances. The circumstances are characterized by common problems confronting an entire branch of industry and thereby demanding joint action. Therefore, trade associations have been the natural starting point for co-ordination of the problem-solving efforts. However, the trade associations cannot put their members under legal obligations. This means that the agreements have to be voluntary and non-binding in a legal sense.

With regard to environmental effectiveness, it has been found that in many fields, agreements only codify a status of behaviour which has already started at the enterprises in question rather than initiate a new development. In these cases environmental agreements do not seem to possess the same environmental potential as economic means, because they lack a continuous incentive to optimize and innovate.

The general impression concerning the field of waste and recycling is that agreements are effective when the problem addressed is relatively simple. When more complex connections are concerned and the development of better returnable products has been required, the agreements do not seem to have been sufficiently effective.

In connection with several of the agreements concerning return and recycling schemes, a system of fees or taxes has been imposed. These types of intervention seem to function well, even though the level of the fees and/or taxes naturally influences the results. Discussions on more interplay between agreements and economic means have recently been held.

In relation to administrative costs, the specific measures in the agreements have often been so difficult to agree on that these costs have increased rather than decreased.

The existence of an environmental agreement is in general terms not subject to much public attention. Even amongst professionals working on environmental problems, the knowledge about the existing agreements is rather limited. Although no important NGO generally refuses the use of an environmental agreement, they would have preferred some kind of legal regulation in some specific cases instead. The industry is quite content with the use of environmental agreements instead of traditional legal regulations and even instead of economic measures. They see the agreements as the most flexible and thereby most efficient and gentle measure.

Apart from three agreements in the making, there are no current plans for further agreements. Some divisions of the Danish EPA have decided not to conclude any new agreements before the existing ones have been evaluated.

France

Environmental agreements early appeared in France: They were launched at the beginning of the seventies upon the initiative of the newly established Ministry of the Environment, which wanted to demonstrate quickly its ability and efficiency in dealing with pollution problems. Since French public authorities already had quite a good relationship with industry, they preferred to work 'hand in hand' with the industrial world rather than having to develop new command and control means.

The first environmental agreements were named 'contrats de branche' and were concluded with various sectors such as the paper, sugar and leather industries. They provided the allocation of financial subsidies by the state to the private partners.

Due to the possible infringement of the European Treaty provisions on state aid by these 'contracts', the government quickly reshaped its policy on environmental agreements by separating them from subsidies. The new agreements were named 'programmes de branche/d'entreprise'. They involved the groups 'Pechiney-Ugine-Kuhlman' and 'Creusot-Loire', for instance.

In the late seventies, a case was brought before the French Conseil d'État that challenged the authority of the Minister of the Environment to enter into an Environmental agreement. The Court stated that an authority cannot act via a contract when empowered to act via classic regulatory or policy means.

Recent but outdated agreements dealt, among other things, with beverage packaging, CFCs, and clean-up of land. Existing environmental agreements today include:
- an agreement on end-of-life vehicles;
- various agreements on greenhouse gases reduction (as an alternative to CO_2 taxes);
- agreements on waste management.

The legality and binding force of current environmental agreements is not straightforward. Information on most environmental agreements is difficult to get. These documents are not published, nor are they made easily available to the public.

The scale of the territory and the current trend towards decentralization of competencies in France are factors that affect negotiating processes. Nowadays, negotiations often take place between various public authorities or between local public authorities and private persons in order to deal with environmental issues which lead to the conclusion of environmental agreements.

Germany
In the German environmental policy environmental agreements with at least one party from industry have become very popular. On the one hand, this is a consequence of the federal government's desire to intensify the use of co-operative instruments such as agreements in their policy and to reduce governmental legislative intervention. On the other hand, since the late 1980s, associations of different industry branches issued a large number of new agreements concerning various environmental issues.

Most of the almost 90 agreements that were identified on a national or state level in the German Country Report are so-called self-commitments (Selbstverpflichtungen) or voluntary/negotiated agreements (freiwillige Vereinbarungen). It is not possible to derive agreements strictly from self-

commitments. Both terms are used alternatively, even when people are talking about the same agreement. When environmental agreements are classified in relation to their environmental target, they mostly fall under one of three categories:
- Phase-out obligations (related to the use of certain substances, such as CFC, asbestos etc.);
- Reduction targeting obligations (relating to emissions or the use of resources such as the climate protection self-commitment of industry);
- Take-back and recycling obligations (e.g. bottles, batteries, old cars).

The vast majority of environmental agreements in Germany have some common features:
1. The abstention from one-sided governmental constraint and the activation of private initiative. The state generally holds out a prospect of exemption from (new) regulations. The grounds for exemption are announced explicitly or are understood by implication. In some cases the government provided financial assistance or supporting measures in compliance with the wishes of the industry in environmental related or other issues.
2. The exclusion of third parties. Negotiated agreements and self-commitments of the past have almost always been concluded without any involvement of third parties such as environmental or consumer organizations.
3. The non-binding character of the obligations. In fact, agreements generally contain only a moral obligation of the participating parties to abide by their promises. They are not legally bound to comply.

Particularly these three characteristics have led to a rather low acceptance of this instrument by the environmental movement in Germany. The main arguments of industry and policy-makers in favour of environmental agreements is that they are better suited to the market economy than legislative measures and would appeal to industry to take responsibility for its own affairs. However, it is almost impossible to evaluate the economic aspects, since data are either not known or hardly available.

The experience with environmental agreements in the past regarding their environmental effectiveness was quite different. A preliminary evaluation shows that phase-out obligations often reached the targets while many other agreements failed for various reasons. Specific reduction targets, for example, that were achieved have been overcompensated due to increased production figures. As a result, the total environmental effect was negative in the end.

More and more industrial branches are announcing new self-commitment initiatives, and policy-makers are still in favour of environmental agreements. In that light, the significance of this instrument for environmental policy in Germany is expected to rise in the future.

United Kingdom

The history of the modern environmental agreement in U.K. environmental policy is short and has two distinct policy contexts. The first is deregulation (and its links with competitiveness). The second is a general trend towards the promotion of 'partnership' approaches in formulating and implementing environmental policy. There is no formal policy framework for the adoption of environmental agreements (understood as government-industry negotiated agreements) at the national level. Nor is there any dedicated legal framework for the negotiation of government-industry agreements.

The first government-industry environmental agreements that can be readily cited in the context of contemporary policy debate about the role of environmental agreements were concluded in 1996 between the Department of the Environment and a series of trade associations. These five non-legally-binding agreements address HFC use. Only three contain any commitment on reporting. A fourth provides for the air-conditioning and refrigeration industry to 'assist' the government in obtaining regular information on HFC use and emissions. None of the HFC agreements contain any provisions on monitoring; only three incorporate provisions for review, and there was no public participation in the negotiation of any of the agreements.

A 'producer responsibility' initiative has also produced strategies that fall within the scope of the broad term 'environmental agreement'. The way in which the principle of 'producer responsibility' has been implemented in the U.K. has in some cases resulted in commitments that are reminiscent of German 'self-commitments'. Many of the waste streams that have been singled out within the producer responsibility initiative (batteries, packaging, tyres, end-of-life vehicles, electronic equipment) are linked with EC legislative initiatives. However, there are also examples (e.g., packaging and newspapers) of an express threat of legislation providing an incentive to spur on business-led voluntary approaches. In response to the 'free rider' difficulties experienced in relation to the packaging producer responsibility initiative, the Environment Act 1995 introduced provisions enabling the Secretary of State to make 'producer responsibility' regulations. So far, the powers have only been applied in relation to the packaging waste stream.

A primary value in the U.K.'s approach to environmental agreements appears to have been 'flexibility'. But flexibility has often been maintained at the expense of
- quantified targets;
- clear and verifiable monitoring and enforcement mechanisms; and

- transparency of process and implementation.

The new government has expressed cautious interest in the further development of voluntary or negotiated agreements. Ongoing negotiations for an agreement on energy efficiency in the chemical industry will be a key testing ground. Ministers want to ensure that environmental agreements do not simply represent 'business as usual'. There is also a possibility that the newly created Environment Agency could eventually find a role for 'negotiated agreements' in some form when carrying out its regulatory functions - particularly in its approach to companies that have developed Environmental Management Systems.

Italy
The environmental agreements entered into voluntarily represent a new phenomenon for the Italian legal system. The first of these agreements date back to the late eighties. None of the agreements provide for any legal consequences in case of non-fulfilment by one of the parties of the undertakings agreed upon.

Approximately 25% of the agreements inventoried have been concluded as a consequence of a regulatory legislative programme; approximately another 25% concern the waste sector; and the remaining 50% of the agreements cover air pollution (25%), water pollution (one agreement), self-obligation programmes of the agreements (four agreements), and Neighbourhood Agreements between NGOs and commercial sectors (three agreements).

As in other countries, environmental agreements in Italy often aim at integrating (and in some cases replacing) traditional public administration law or common law.

The role of agreements in environmental policy is nevertheless limited. To be valid, environmental agreements must not conflict with mandatory rules. Moreover, the majority of obligations imposed by environmental law in Italy are accompanied by a penal sanction. This requirement limits the applicability of environmental agreements considerably; they may only guarantee implementation of legislation. They cannot autonomously establish environmental goals different from those pursued by legislation.

The Netherlands
Almost 80 environmental agreements between public bodies and industry have been identified in the country report on the Netherlands (state of affairs August 1996). This list excludes environmental agreements in which industry is not involved as a partner. Four groups of agreements are distinguished:
- declarations of intent in the context of the Dutch target group policy on industry;
- multiyear agreements on energy efficiency in industry;

- product-related agreements;
- other environmental agreements.

The first group is the smallest but the most wellknown, internationally. These agreements are part of a completely new form of environmental policy-making, stimulating partnership and the responsibility of industry. The second group is the largest, but it only relates to energy-efficiency targets. The third and fourth group include the oldest types of agreements, dating back to the early eighties. They usually address specific environmental problems which are sometimes difficult to solve efficiently by other means than an agreement.

Extensive discussions took place among industry organizations, policy-makers, politicians and societal groups in the late eighties and the start of the nineties on the merits of environmental agreements. They are now more or less accepted, mainly due to the lessons learned from past mistakes. Very important, for example, is the increased transparency in the implementation results of the agreements, also on the level of individual companies. Through experience, the parties have learned that an agreement usually cannot be a useful policy instrument in itself. It needs support from other instruments, notably the permit system, in order to be effective and to deal with free riders. On the basis of advice from several legal bodies and a political debate, in 1996 the Dutch government issued generally applicable guidelines for covenants. The Dutch debate is now concentrating on what to expect from the instrument in the future (2000-2010). For example, the parties are considering what influence environmental agreements can have on the establishment of tradable permit systems for NO_x.

A preliminary evaluation of the environmental effectiveness of the environmental agreements shows a rather variable picture. Depending on the subject and the partners in the agreement and many other parameters, some agreements are or seem to become effective while others are not. Economic efficiency is not an issue at all in the Netherlands and data are not available. Public participation is mainly limited to a passive role of the public through the availability of all kinds of monitoring, evaluation, and progress reports. The social acceptance is in general rather high.

Poland

The situation in Poland regarding the use of environmental agreements in environmental policy has the following characteristic features:

- There is insufficient legal basis for the use of environmental agreements in public environmental policy.

In Poland, constitutional arrangements require public officials to act exclusively in accordance with the legal procedures provided for the purpose. Traditional civil contracts are sufficient for 'Neighbourhood Agreements' but are not considered to be appropriate to replace command and control instruments in the relationship between regulators and the regulated community.

- Environmental 'covenants' (i.e., agreements between public administration and industry associations) do not yet exist.

The entire idea of 'covenants' is relatively new. Traditionally, instruments of environmental policy have been designed to induce officials to take a 'hard look' rather than to seek consensus with polluters. The groups that represents the interests of 'polluters' is not strong enough to make such covenants a real alternative to other means to implement environmental policy. Associations which claim to represent the industry are divided internally. They are far too weak to provide safeguards against massive 'free riding'.

- There are various practical arrangements on environmental protection which may be called 'environmental agreements'.

These include agreements between environmental authorities and particular polluters, as well as agreements between polluters and affected people or local communities. In most cases, it is difficult to determine the legal status of the agreement because diverse legal forms have been applied.

- There are no statistical data concerning environmental agreements, and access to government records is insufficient.

There is still no right of access to information. There are no precise procedural rules governing disclosure of government information, data, and documents. This made research very difficult; it had to be based on an arbitrary selection of information.

- Neighbourhood Agreements between polluters and those affected by their pollution have a long tradition in Poland.

Most of these agreements were limited to paying compensation while neglecting the need to abate pollution. Only recently has a new tendency emerged towards the conclusion of Neighbourhood Agreements oriented to pollution reduction or at least to mitigation measures.

United States

In the U.S., the 1990s have seen a dramatic shift toward greater use of environmental agreements, initiated both by the government and non-governmental parties. One of the first major voluntary arrangements in the U.S. was reached in 1991 when the U.S. Environmental Protection Agency (EPA) acted upon public concern with industrial pollution, a concern prompted by the mandatory federal Toxic Release Inventory, a public 'Right to Know' law database of chemicals emitted by industry to spur a voluntary 'Industrial Toxics Project' generally known as '33-50' . The voluntary approach of '33-50' was a forerunner of a large number of EPA programmes which take a similar voluntary approach. These programmes address energy consumption and various pollutant emission concerns.

In general, there is heightened interest in many circles in a new model of 'partnership' which moves away from the 'adversarial' and 'command and control' regulatory traditions of the U.S. toward collaborative problem-solving.

The President's Council on Sustainable Development, a multi-stakeholder panel convened by President Clinton, issued a report in 1996 entitled *Sustainable America*, which attempts to characterize and foster some of this collaborating work. Another important arena for environmental agreements is the area of regulatory reform. There are extensive activities in the U.S. to promote regulatory reform, built around environmental agreements, in which the *quid pro quo* for regulatory or enforcement relief is cleaner performance and new forms of accountability. The voluntary arrangements in the U.S. have typically emerged in a context in which other litigious, legislative, and regulatory tools are available in the background as potential sources of recourse and 'alternatives to a negotiated agreement.' Some of the farthest-reaching agreements have involved creative interfaces with non-voluntary proceedings: permitting under air pollution, water pollution or hazardous waste regulations; land use approvals. Indeed, lawsuits to seek enforcement regarding violations of environmental laws, -some brought by government prosecutors and some by non-governmental organizations-, have also spurred creative agreements known as 'supplemental environmental projects.'[8]

Ongoing discussions, deliberations, and decision pipelines are likely to lead to an explosion of new voluntary agreements in the U.S. in the late 1990s. Some of the most important future developments in this field may be expected to derive from the EPA's Common Sense Initiative (CSI). The CSI is a broad programme of multi-stakeholder dialogue built around selected industrial sectors: e.g., petroleum refining, printing, computers and electronics, iron and steel, automobile manufacturing, and metal finishing. CSI's multi-stakeholder teams have forged consensus on nearly 40 projects to test innovative approaches to public health and environmental protection. These include projects to:
- Reduce duplicative reporting requirements;
- Streamline the permits process;
- Improve community involvement in environmental decision-making;
- Find incentives for and eliminate barriers to pollution prevention; and

---

[8] USEPA, Office of Regulatory Enforcement: Innovations in Compliance and Enforcement: Supplemental Environmental Projects in EPA's Toxics and Pesticides Program: September 1994; USEPA, Interim Revised Policy on the Use of Supplemental Environmental Projects in EPA Enforcement Settlements, May 3, 1995.

- Explore alternatives to the current regulatory system to provide more flexibility in meeting environmental standards.

Another experimental programme which may show eventual results is EPA's Environmental Leadership programme. It aims to test a variety of environmental compliance management systems as alternatives to traditional enforcement approaches. The companies who enter ELP pilot project agreements with the EPA expect co-operative enforcement oversight in exchange for a commitment to higher levels of disclosure and accountability. Other trends include the following:
- Several states, including New Jersey, are in the process of developing programmes loosely modelled after the European "Green Plans" as an alternative to traditional regulatory approaches.
- Project XL, a case-by-case regulatory reinvention programme of the EPA, has numerous project applicants and participants in its decision pipeline.
- Environmental and social justice organizations, and organized labour, have been launching new initiatives and pressure campaigns geared to encouraging U.S. corporations operating in the U.S., and also outside the U.S. borders, to enter into various forms of 'sustainable development' commitments.

Some of the most important environmental agreements in the US have been reached at the neighbourhood level. These agreements have generally been reached where effective local civic and environmental organizations have sought agreements directly from local companies which engage in pollution or which pose other hazards (e.g., chemical accidents) to their neighbourhoods. These Neighbourhood Agreements are known as Good Neighbor Agreements, or GNAs. These largely by-pass government agencies and officials as signatories. Instead, they emanate from organized pressure that grassroots environmental groups exert directly on local industries.

5.4 Comparative overview

Use in environmental policy sectors
A comparartive look at the application of the environmental agreement in the different countries reveals a preference for specific policy sectors. In all member states of the European Union that were examined, the instrument has been used in the waste management sector. The survey showed that specifically the Netherlands, Germany, France, Denmark, and Italy turned to environmental agreements in order to solve waste problems. The Netherlands, for example, has concluded 17 agreements related to waste management. The popularity of this

instrument for waste management in Germany (16 agreements) and France (12 agreements) is almost as high.

Another policy field in which almost all countries have used environmental agreements is energy and climate change. The greatest number can again be found in the Netherlands and in Germany. France, Denmark, the United Kingdom, and the United States have also used the instrument to implement their climate change policy. Ozone depletion is a field of application in Germany, Belgium, France, and Denmark as well as in the Netherlands.

The prevention of water pollution is also an area where environmental agreements are used. However, the total number of agreements in this sector is much lower. Not all countries examined used this instrument to improve water quality. Germany, the Netherlands, and Denmark proved to make use of this instrument concerning water pollution more than other countries.

A comparatively smaller number of agreements have also been concluded to protect human health and to prevent air pollution. No agreement has been made with regard to soil protection.

It can be concluded that environmental agreements are mostly used in waste management policy and in climate change/energy policy. This preference does not necessarily lead to the conclusion that the instrument is most suitable in these sectors. It might also reflect a preference for certain topics in environmetal policy in general or specific interests of the economic players within the sector.

Significance for environmental policy

Environmental agreements were arranged relatively early in France and Germany. Both countries had their first agreements in the seventies. In France, the first agreements were initiated by the newly established Ministry of the Environment. The officials wanted to demonstrate their ability to act fast and work efficiently. Public authorities traditionally maintained good relations with industry and thus preferred a co-operative approach. Due to legal proceedings of the Conseil d'Etat, the process was slowed down in the eighties. In the nineties, the instrument was revived.

In Germany, the first agreements did not have much significance for environmental policy until the eighties. Between 1984 and 1987, industry announced a significant number of environmental obligations to avoid regulation. In the nineties, the instrument became even more popular. This is due to the coalition agreement of 1994 which gives priority to 'self-commitments' over regulations for the waste sector. The self-commitments emerged in a situation where 'over-regulation' in the environmetal policy sector was often blamed by industry associations for making Germany unattractive to investors. The self-obligations usually were less ambitious than the alternative, namely legislative proposals. The instrument is for this reason quite controversial.

The other countries started their activities considerably later. In the Netherlands, the first agreements were signed in the eighties. However, the application on a large scale started only after publication of the first National Environmental Policy Plan in 1989 and its addendum NEPP+ in 1990. The agreements of the nineties are more carefully designed than the agreements of the first generation and most are part of an overall strategy of the government. A supporting policy is an essential part of the strategy. Of all the countries reviewed here, the Netherlands has concluded the most agreements in the nineties.

A number of other countries also started to make use of agreements as an environmental policy tool in the eighties. However, the instrument does not play an equally important role in each country. Belgium, for example, started to apply the instrument in the late eighties. The division of competence between the regions seems to be an obstacle to strong national agreements. The agreements are thus arranged at the regional level. The Flemish region has passed a decree with a specific framework for environmental agreements. So far, no agreements have been reached under this new law. Negotiations between the government and industry are still carried on in a rather informal matter, with little possibility for the public to gain access to information.

The emergence of voluntary agreements in Denmark in the eighties is the result of an effort to make environmental regulation more efficient by cutting down on bureaucracy and creating more flexibility and self-regulation. These agreements are also the result of pressure from representatives of industry and trade who wanted to reduce the extent of legal regulation. In this respect, similarities to the German debate can be observed. Since 1991, the Danish Environmental Protection Act has contained a specific provision on binding agreements in Section 10, but it has almost never been put to use. In this respect, parallels to the attempt in Belgium can be drawn. Possibilities for more interaction between agreements and economic means have recently been discussed. The new Danish CO_2 tax system contains elements of voluntary action, because in specific cases voluntary activities can lower the tax rates. However, the number of agreements between government and industry is low and the instrument has not gained as much significance as in the Netherlands and Germany. Some divisions of the Danish EPA have decided to evaluate the agreements up till now before any new agreements can be negotiated. This might reflect a certain scepticism regarding the effectiveness and efficiency of the instrument.

The history of the instrument in the United Kingdom is comparatively short. The emergence of voluntary agreements as a distinct environmental policy tool in the U.K. has to be seen in the context of an increasing concern about the ability of law to deliver environmental objectives - particularly in the face of increased understanding of the complexity and diversity of the sources of environmental problems. The new Labour government seems to carry on the approach, though it emphasizes that agreements should not lead to 'business as usual' but to improved environmental protection.

The first agreements in Italy were only made in the late eighties, but the instrument became significant in the nineties. In 1988, the Italian Legislature set up the *Consorzi Obbligatori* for recycling. The regulatory programme provides for mandatory co-operation between sector companies in order to attain a specific environmental goal. Apart from the waste management sector, environmental agreements seem to be a less common instrument than in the other European Union member states examined here.

The policy context in the countries that are not member states of the European Union -Poland and the United States- shows some models and motives for environmental agreements.

In Poland, agreements between government and industry associations have not been made. Instead, agreements are used between authorities and single polluters in order to reduce pollution. The 'Top 80 polutters programme' encouraged the polluters to prepare a pollution reduction programme and offered special conditions. For instance, it set very liberal conditions on the environmental permits for the duration of the programme and provided subsidies. The action has to be seen in the context of the practical need to reach compliance with new standards in the fastest way possible.

In the U.S., the 1990s have been a time of experimentation with environmental agreements and consensus-based processes, initiated both by the government and non-governmental parties. One of the most prominent U.S. approaches is the 33-50 programme. Under that programme, single polluters are encouraged to commit themselves to the reduction goal that was set up by the Environmental Protection Agency. The driving force behind 'environmental agreements' nowadays, like in many European Countries, is the search for alternatives to command and control measures for more efficient policy tools.

5.5 Conclusions

Environmental agreements have been used as a policy tool in the seventies in France and Germany and in the early eighties in the Netherlands. However, they were not expected to play a significant role in environmental policy.

By the end of the eighties and in the early nineties, the popularity of this instrument had increased considerably. The countries that already had some experience started to work on a second generation of agreements. All other countries examined here -Italy, Belgium, Denmark, the United States, and Poland- turned to this instrument too.

The increased use of environmental agreements was mainly motivated by scepticism towards conmmand and control measures and the hope for more efficiency and flexibility. This applies especially to Germany, Denmark, Belgium, and the United Kingdom and recently also to the United States. Additionally, environmental policy seems to turn to more co-operative approaches. This shift is most evident in the Netherlands and the United Kingdom, but it was also detected in the other countries examined here.

It can be observed that environmental agreements in the member states of the European Union are usually arranged between industry associations and the government. In the two non-member states that were examined -Poland and the United States- the agreements were signed by the government or a governmental authority and single companies, on the initiative of the government.

In the countries that have attempted to establish a legal framework for environmental agreements -Denmark and Belgium- this framework was not used. This suggests that a certain degree of informality is necessary in order to hammer out an agreement.

All member states had agreements in the waste management sector. Almost all member states applied the instrument in the sector of climate change/energy. In relation to other policy fields, the picture is more diverse.

All in all, the comparative overview shows that in spite of the diversity with regard to legal form of and integration into the legal system, some similar practices can be observed among the member states. This applies to the motives, the relevance to environmental policy, and the policy fields in which it is mainly applied.

The following chapters provide a more thorough analysis of environmental agreements in several countries. That analysis considers the agreements from various disciplinary angles and looks at this instrument from a range of theoretical perspectives.

6. SUCCESS DETERMINING FACTORS FOR NEGOTIATED AGREE-MENTS

A comparative case study of the Belgian electricity supply industry and the packaging sector

AKIM SEYAD, STEVEN BAEKE, MARC DE CLERCQ

6.1 Introduction

Environmental issues have to be tackled within a network of public and private actors. However, traditional instruments of environmental policy were not always designed to deal with the multifaceted, self-referential, interdependent, and dynamic nature of this network. Therefore, most governments have come to acknowledge that an environmental policy based solely on unilateral regulation is no longer adequate. Thus, there is a need for a new way of making environmental policy making. The new approach must be based on a mix of traditional and 'novel' instruments of environmental policy. The new measures must be more flexible and more appropriate to the characteristics of a network.

The negotiated agreement is one of these 'novel' instruments. It is an agreement between industry and the government wherein industry promises to achieve certain environmental goals whereas the government commits itself not to issue any regulation, during a specified time span, that might overlap with the content of the negotiated agreement.

The basic idea we wish to explore here is that the economic-institutional context in which a negotiated agreement is used strongly influences the performance of that agreement. Therefore, in order to fully exploit the potential of negotiated agreements, the parties have to know which factors of the institutional context have a positive effect on this performance.

The objective of this paper is to evaluate the performance of negotiated agreements in Belgium's electricity supply industry (ESI) and in the packaging sector. In order to draw conclusions, we must determine which factors within the economic-institutional context of those sectors now have or once had an influence on this performance. That is done through a comparative case study analysis. The following hypotheses could be formulated with regard to the link between the institutional setting and the effectiveness of voluntary agreements:

P. Glasbergen (ed.), Co-operative Environmental Governance, 111–132.
© 1998 *Kluwer Academic Publishers. Printed in the Netherlands.*

a) Negotiated agreements are more easily concluded and will be more correctly observed if there is already a tradition of joint policy-making between the government and the sectors concerned.

b) Negotiated agreements need a common negotiating position for industry. Such a position can be compared to the production of a collective good. It is more easily achieved if the number of participants is limited and/or the problem of free riders can be controlled.

c) Industry will be more likely to comply if it has a real economic advantage (for instance, lower compliance cost) in the voluntary agreement regime, an advantage that would not be guaranteed with other instruments.

d) The effectiveness of the agreement is better assured if an effective monitoring system is in place.

e) Industry will show more respect for the measures taken when the government seems determined to use more painful instruments if the agreement does not work.

f) It is easier to reach agreement if there is consensus on the underlying objectives. This means that the policy goals are considered relevant by the industry concerned and that they do not conflict with its own strategic objectives.

By comparing the voluntary agreement in Belgium's electricity sector with that reached in the country's packaging sector, we were able to draw some tentative conclusions with regard to these hypotheses. To follow the argument, the reader should have an understanding of the institutional setting of the country. Therefore, the next section will highlight the relevant structures and how they are used.

6.2 Belgian public policy

General public policy: structure and style
The whole political system in Belgium was altered by the Federal evolution law of 1993. Belgium is now a federal state composed of three regions (the Walloon, Brussels, and Flemish region) and three communities (the Flemish, French, and German community). The regions mainly deal with territorial matters (environment, economy, transport, …), while the communities deal with person-related matters (culture, education, …). Belgium has its own brand of social dialogue and compromise. That system is formally organized in councils and committees. The main social pressure groups have a direct impact on government policy through their influence on the selection of candidates for the Belgian parliament. Especially trade unions have a strong impact on the political parties. They also have influence through their participation in the political cabinets of ministers and through their seats in several consultation councils and even in

decision-making bodies (e.g. SERV, MINACouncil). On the federal level, for example, there is a national council for sustainable development, composed of the main social groups of the country (employers associations, trade unions, farmers, consumers, etc.).

Public environmental policy
In Belgium, environmental policy is mainly dealt with by the three regional governments of Flanders, Brussels, and the Walloon region. The federal government is left with only limited powers in this field. Its purview covers product standards, nuclear waste, and the negotiation and implementation of international commitments (e.g., the introduction of EC directives in Belgian environmental law). It is important to note that there is no hierarchy of legal systems between the Federal State and the regions; each has its own constitutionally guaranteed competencies. Hence the national government can not impose its will on regional governments in environmental matters. In cases where they are all stakeholders, the national government and the regional governments have to consult with each other in order to negotiate a common position. Prior to the decisions of the Flemish government on environmental policy, advice is rendered by the so called MINACouncil. This is a consultation council, composed mainly of representatives of environmental movements; the social and economic groups are represented in a minority position. In environmental matters, advice is also given by the Social and Economic Council of the Flemish Community (SERV), composed of representatives of the main social groups and of scientific experts (a limited number of members). This advisory structure has a severe disadvantage. In most cases the MINACouncil (or at least a majority of its representatives of environmental groups) takes almost exclusively environmental motives into consideration, while the socio-economic considerations are provided for through the SERV. As a result, real integration of the economy and the environment does not happen on the consultative level and has to be done by the policy-maker himself without real preparation.

Environmental policy is characterized by command and control regulations that are grouped in permits. Till recently, the regulatory system was rather complex and the sanctions low, so compliance was not at all guaranteed. Recently, the Flemish government appointed a commission that proposed a systematic review of the environmental law of Flanders. As a consequence, the control and sanction system was considerably upgraded. The Flemish government recently announced the installation of a special commission that will investigate the effectiveness and the efficiency of environmental laws.

Although negotiated agreements are recognized as an instrument of environmental policy in Belgium, they are not systematically applied, as in the Netherlands. Their use depends strongly upon the goodwill of the regional ministers of the environment. Until June 15, 1994 approximately 14 negotiated agreements were signed with different industry federations. However, despite the lack of a systematic approach, Flanders is probably the only state in Europe which has instated a legal framework (June 15, 1994) for negotiated agreements (see further). This legal framework was intended to create a context wherein the use of negotiated agreements would be more successful.

As stated above, environmental issues fall mainly under the jurisdiction of the regional governments. However, we will not go into the Walloon or Brussels' environmental policy, as the structure is rather analogous to the Flemish structure.

The Flemish decree on negotiated agreements
In 1993-1994 a study on negotiated agreements in Belgium was performed by the Centre for Environmental Law at the University of Ghent. This study revealed the following facts:
- Many industry federations, together with their members, have elaborated an environmental policy for the sector. However, some juridical problems arise when a federation signs a negotiated agreement representing its members.
- A simple regulation of an authority, not necessarily the contractor, can have a negative influence on the outcomes of the negotiated agreement.
- The characterization of the legal statute of a negotiated agreement by different parties is often contradictory in the sense that one party talks about a binding agreement whereas another party refers to the arrangement as a gentlemen's agreement.
These findings apply to all Belgian environmental agreements that were signed before July 8, 1994, the date on which the Flemish decree on negotiated agreements was published in the official state journal. This decree was instated in order to create a legal framework to adress the problems mentioned above. Unfortunately, no negotiated agreements have been signed since that date, which makes it impossible to evaluate this legal framework. However, it is believed that this framework itself is responsible for the fact that no negotiated agreements have been signed since the implementation of the decree. The main reason for this is that the unrealistic and long procedure foreseen in the decree does away with many advantages (like flexibility, speed, ...) of negotiated agreements, which makes this instrument less attractive.

Although the legal framework does not apply to the negotiated agreements of the cases studied here. It is still interesting to consider it here, as it is unique in Europe. The following elements are covered by the decree:
- The decree defines the concept of negotiated agreements.

- The decree defines the parties that can sign negotiated agreements, namely the Flemish region (government) and one or more representative industry organization who have to fulfil certain requirements (corporate entity, representative of companies with similar activities or environmental problems and mandated by these companies).
- The decree defines the procedure that has to be followed: publication of the draft negotiated agreement in the official state journal; with the right of general appeal during 30 days; non-binding advice of the SERV and of the MINACouncil within 30 days; final decision by the Flemish council within 45 days following the advice and appeal period; publication in the official state journal.
- The decree states that a negotiated agreement cannot replace nor be less stringent than existing regulation.
- The decree states that the Flemish region cannot issue any regulations that are more stringent than the negotiated agreement (except in case of urgency or in order to meet the international or European obligations of the country). The Flemish region is still allowed to put the content of the negotiated agreement into regulation.
- The decree states that the negotiated agreement is binding for the representative organisation and its members. (Members that adhere to the representative organization are bound and members that withdraw from the representative organization remain bound by the negotiated agreement.)
- The decree states that the duration of the negotiated agreement can be no longer than two years.

6.3 Case study of the electricity supply industry (ESI)

Environmental impact of the electricity sector
Just like other human activities, the production of electricity in Belgium has a major impact on the environment. The environmental impact starts with the use of non-renewable natural resources (gas, oil, and coal). After that, during the production of electricity, NO_x and SO_2 emissions are released into the air. Belgium also has the problem of nuclear waste, which is generated in the two nuclear reactor parks of Doel (near the Dutch border) and Tihange (near the French border). Particularly relevant are the environmental effects that occur during the production of electricity, namely the greenhouse effect and the acid rain problem. These environmental problems have been put high on the environmental agenda. It is commonly acknowledged that they are caused by respectively NO_x and SO_2 emissions. Forty percent of the Belgian electricity comes from the combustion of coal, oil, natural gas, landfill gas, solid waste, and sewage sludge. The rest of the electricity is produced in nuclear plants.

The environmental consequences of *coal- and oil-fired* power plants include emissions of NO_x and SO_2 as well as particulates, toxins, and incompletely burnt hydrocarbons. As electricity is generated largely in steam turbines, another consequence is thermal effluent in the cooling water, with effects on the water quality and the aquatic habitat.

The primary problem with the combustion of *natural gas* is control of NO_x emissions. In general, natural gas emits less NO_x per unit of electricity than oil or coal. In comparison with oil and coal, the environmental performance of natural gas may favour the use of gas in electricity generation in three ways. First, the costs of meeting existing regulations for SO_2 and NO_x are generally lowest for natural gas. Second, over the lifetime of a plant, natural gas use is less likely to be affected by tighter environmental standards, since it is a relatively clean fuel. Third, natural gas can be used in dual-, multi-, or co-firing combinations as part of a strategy to meet emission limits.

The problems associated with energy derived from *municipal solid waste* are determined by the chemical make-up of the waste, its transport, and the conversion process. Typically, this waste contains large amounts of hard glass and metal, including heavy metals.

Combustion of *landfill gas* is considered environmentally beneficial. Methane, which contributes significantly to the greenhouse effect, is converted into CO_2, which has a weaker greenhouse effect. Though SO_2 emissions are higher, emissions of other pollutants from the small gas turbines that run on landfill gas are comparable to those of natural gas plants.

The problems associated with electricity derived from the combustion of *sewage sludge* result from microbes, organic sulphur, and traces of hazardous chemicals and heavy metals.

Market structure of the electricity supply industry
At the base, the economic context of the ESI was modelled by a legal framework, formalized in 1925. The laws granted communes the right to determine how supply would be met in their area. The industry was also shaped by the influence in the Belgian economy of industrial holding companies, such as Société Generale Belgique (SGB) and Bruxelles Lambert. Their dominance has ensured a major role for privately owned production companies in the sector. The mixture of predominantly privately owned producers and publicly owned or controlled distributors has continued ever since.

There was pressure to bring the sector into public ownership in the 1950. It was felt that the development of the ESI was a prerequisite for postwar reconstruction and was best guaranteed by state control. However, this pressure was resisted. The industry was subjected to an institutionalized form of regulation and scrutiny. A control committee was installed in 1955 to regulate prices and investments. In 1964, it became the Comité de Control de l'Electricité et du Gaz (CCEG).

The Belgian energy sector in general and the power sector in particular have undergone drastic change in the last few years. In 1990, the three private utilities (EBES, Intercom, and Unerg) merged to form one private company: Electrabel. According to the government, the objectives of the merger was to create an entity better able to resist possible take-over and with a better competitive position in the emerging European single energy market. Since the closure of Belgium's last coal mine in 1992, the country imports nearly all its primary energy needs.

Electrabel dominates the power generation sector. It has significant interests in the distribution of electricity and gas through numerous mixed low-voltage distribution companies it owns jointly with the municipalities. In addition, there is a public power-generating company, SPE.

The market strength of Electrabel is enormous, and the regulatory response has not been correspondingly strong. As stated above, government regulation is done through the Control Committee on Electricity and Gas and the National Energy Committee. The CCEG consists of members from the employers' federation and the trade unions. However, it has very limited resources and only advises the government regarding cost price trends, the organization of the sector, investment plans, and tariffs.

About 94% of the country's total electricity demand is met from power stations owned by Electrabel. The balance comes from SPE and autoproducers. Tractebel, the large conglomerate that effectively controls Electrabel, is majority-owned by the mixed private group Société Generale Belgique (65%).

Distribution is based in and controlled by the approximately 600 municipalities, which have a monopoly on low-voltage distribution. High-voltage customers can buy directly from Electrabel; about 65% of all electricity is sold this way. The distribution business has been characterized by an increasing degree of concentration. Currently, there are 20 companies of mixed ownership, with Electrabel being the largest private investor; 14 municipal companies; and nine fully public-sector distributors. SPE is publicly owned and has a complex ownership pattern. The main shareholders are municipal and regional electricity distributors, plus several state-owned financial companies and a few private investors. This fragmented ownership pattern means that Electrabel has been able to point to the risk of a foreign take-over as a justification for wishing to take over SPE itself.

The significant degree of cross-ownership in the industry -together with a high degree of vertical and horizontal integration, a light form of regulation, and a mixture of formal and informal relationships among production, transmission, and distribution companies- has produced a centralized and monopolistic power sector. Competition is carefully avoided, the barriers to entry of new generators are relatively high, and consumers' choices are limited. It comes as no surprise that electricity tariffs in Belgium are among the highest in Europe.

The various levels of government have been able to influence general strategies in the energy sector via shareholdings in selected companies, semi-official bodies, and agreements with industry. However, there are concerns about the uncertainties created over R&D and environmental policy by the fragmentation resulting from federalization. A ten-year plan for developing generating resources has been agreed between the federal government and the ESI, and the CCEG has commented on it.

There is no formal legal restriction on the production or sale of electricity, though the conditions of sale have to be negotiated. Tariffs are related to avoided costs, and the buy-back price structure is reviewed by the Ministry of Economic Affairs as well as by the CCEG.

Public regulation
The ESI is highly regulated by the government. Government regulation of the electricity supply industry (ESI), which is carried out by the Control Committee and the National Energy Committee, covers tariffs and investment plans. The basic framework was set up in the 1950s and was expanded by the 1980/81 reforms. The Control Committee is a typically Belgian coalition involving members from the employers' federation and the trade unions whose representatives are carefully balanced between linguistic (Flemish and Walloon) and political (Christian Democrat and Socialist) interests. This committee regulates the electricity producers (Electrabel and SPE), as represented by the CFEE, and the distributors' associations (Intermixt and Interregies).

The secretariat of the committee is four-strong, drawn from the employers' federation and unions, again with a mix of ethnic and political representatives. Unlike the US public utility commissions, but like most European regulators, the permanent staff of the committee is minimal. As a result, the committee has rather limited competencies. Basically it has two tasks: monitoring cost price trends and profitability; and making recommendations on the organization of the sector, investment plans, and tariffs. The government is not active on the committee, but it can take initiatives by asking the committee to consider a particular option. However, the committee has only an advisory role. The government can, on the basis of broader considerations of national interest, ignore its recommendations.

Price regulation is basically carried out on rate-of-return basis. There is a single set of tariffs which apply nation-wide regardless of distributor. The regulatory procedure is that the utilities submit their financial results to the committee's accountants, who assess their costs. Prices are set according to a formula agreed by the committee on the basis of that analysis. Terms to cover all the major costs including fuel, labour, and financing, are included in this formula.

Indices for movements in each of these terms are derived and consumer processes are adjusted on a monthly basis. This formula is generally valid for a number of years. It comes up for review when there is a significant change in the structure of the generating stock, for example, the commissioning of a nuclear power plant.

Investment regulation is somewhat more complex. The CGEE is required to submit a 10-year investment plan to the Control Committee and the National Energy Committee. That document details intended investments in new plants, major investments in existing plants (e.g. conversion to coal-firing), plants to be decommissioned, and major transmission network investments. These are backed up by electricity demand forecasts. In the first years of operation of this procedure (from 1980 onwards), the CGEE submitted its investment plan annually (1981-83). But in recent years, only two investment plans have been drawn up - in 1985 and 1988. The Control Committee and the National Energy Committee each have 30 days to express their opinions on the plan. If at the end of this period the government has not voiced any opposition to it, the plan is accepted.

At a parliamentary level, there is no standing committee on energy, although there are occasional ad hoc research committees. The standing committee on industry and the economy does carry out inquiries on electricity, but these are infrequent and there is little accumulated expertise. The volatile nature of Belgian politics means that the role of parliament is greater than in other countries and that debates can influence the broad parameters of policy accordingly. Thus, energy policy in the early 1980s was determined mainly by a series of parliamentary debates.

The Belgian electricity covenant

The Belgian electricity covenant was signed on October 18, 1991. The parties on the one side of the table were the Belgian state, the Brussels region, the Flemish region, and the Walloon region. On the other side were Electrabel and SPE. The motivation to forge an agreement was the importance that international, European, national, and regional authorities attached to a policy for the reduction of acidifying emissions. Moreover, the electricity generators recognized that they made a significant contribution to these emissions. But they were also aware that the technological means were available to limit and decrease these emissions, provided that the necessary technological investments could be made in a stable regulatory framework.

It is interesting to note why this stable regulatory framework was created by a voluntary agreement. The reason is that this could only be realized through consultation and dialogue and through a policy instrument that was adapted to the rapidly evolving technological environment. At that time, voluntary agreements had proved successful in the Netherlands and seemed to fulfil the necessary conditions to meet this goal.

The main goal of this voluntary agreement is to reduce the SO_2 and NO_x emissions generated by the installations of the electricity producers mentioned above. The agreement applies to the period up until 2004. That is, it will automatically end on December 31, 2003. However, both parties have agreed to start the negotiations, as of 2001, on a new agreement that would be in force starting from January 1, 2004. The obligations for the electricity producers can be divided into two types. There are general obligations, falling within the bounds of their environmental policy, and specific obligations with regard to overall reductions of their SO_2 and NO_x emissions.

The general obligations can be summed up as follows:
- Power stations must be supplied exclusively with fuel having a low sulphur content (max. 1% S).
- Measures must be implemented to limit the amount of NO_x emitted by combustion. Such measures are decided within the framework of a study programme entitled 'NO_x control' launched in April 1989 by Laborelec (National Laboratory of Electricity Producers).
- Equipment for continuously measuring SO_2 and NO_x emissions must be installed on all large generation units.
- A specific research project in the area of fuel gas cleaning (desulphurization prototype) must be continued.
- It is stipulated that the electricity producers shall participate in governmental development programmes aimed at disseminating environmental technologies in Belgium and abroad, especially in Eastern Europe.

Besides these general obligations, the electricity producers undertake to reduce total SO_2 and NO_x emissions from existing and future power plants. These emission reductions shall be expressed in the form of a comparison with the emissions from 1980, which is taken as a reference year in accordance with the EEC decision on this matter (SO_2: 1993: -70%; 1998: -75%; 2003: -80%; No_x: 1993: -30%; 1998: -40%; 2003: -40%)

The obligations incumbent upon the authorities concern maximum emissions standards for SO_2 and NO_x which may be applied to existing and new generation units, either via legislation or within the framework of operating licences.

The authorities undertake not to impose standards on the combustion units concerned, which are more stringent than those applicable at the time the environmental agreement is signed. Failing that, the agreement shall specify the standards accepted by both parties.

The agreement also stipulates that the authorities may be exempted from these principles in the event that an EU directive is passed which is more restrictive than the provisions laid down in the environmental agreement. The electricity producers shall draw up a detailed annual report of their SO_2 and NO_x emissions. The authorities shall be provided with all information needed to make an appropriate inspection.

A follow-up commission was installed. The majority of its members are representatives of the authorities and it is chaired by one of these representatives. This follow-up commission has the task of ensuring the correct execution of the voluntary agreement and to submit reports to this effect.

Effectiveness of the electricity convenant
The assessment of the effectiveness of a negotiated agreement has to consider whether the goals have been attained in time but should also take the degree of ambition of those goals into account. With respect to the effectiveness, we can draw several conclusions. For NO_x, the goals set in the negotiated agreement have been more than adequately reached, thanks to special burning techniques (the 1998 target was almost met in 1994). In the future, catalytic equipment will denitrify the fumes of modern conventional units. We also see that for SO_2, the goals set in the negotiated agreement have been more than satisfactorily reached, thanks to the more widespread use of natural gas and other low-sulphur fuels and through the energy output of production units (STAG, cogeneration) (the target for 1998 was already met in 1994). In the future, fumes from modern conventional units will be desulphurized. For NO_x, the goals are the same as the European Union targets, while the goals for SO_2 are more ambitious than the European Union targets. In light of the above facts, we can conclude that this negotiated agreement is effective.

Efficiency of the electricity convenant
Efficiency refers to the optimality of the goals set and to the extent to which the instrument has enabled a more cost-effective achievement of policy objectives, compared to some alternative instrument. There is no information available with respect to the cost/benefit of the targets set. No quantitative evaluation is available on the cost effectiveness of the policy either. However some qualitative statements can be made. One can conclude that the instrument of voluntary agreement, has given Electrabel and SPE some freedom to chose ways to deal with the requested emission reductions. Furthermore, it has enabled Electrabel and SPE to plan long-term investments within a stable legal framework. That framework makes it easier for them to evaluate the profitability of these investments. For the government, we can conclude that the monitoring cost of NO_x and SO_2 emissions has decreased, as Electrabel and SPE had to install a continuous measurement programme.

However, it is very difficult to evaluate the efficiency of this negotiated agreement with respect to other instruments of environmental policy that could have been used in this field.

Transparency
As part of the negotiated agreement, a follow-up commission constituted of several stakeholders was installed to evaluate the development of the NO_x and SO_2 emissions. The evaluation of the emissions is based upon the annual report of the electricity sector to the follow-up commission. The status of NO_x and SO_2 emissions is also reported yearly by Electrabel in their environmental report, which is available to everyone.

6.4 Case study of the packaging sector

The problem of packaging waste is rather complex. On the one hand, packaging is necessary. Transportation, protection from damage, preservation of quality and hygiene, and commercialization are some of the functions of packaging. On the other hand, this sector generates a major environmental problem: an enormous amount of packaging waste. This contradiction between the functionality of packaging and the environmental problem it creates has made the packaging waste problem a difficult one to solve. Since the Belgian State Reform of 1980, the Belgian regions have gained almost exclusive competence for environmental policy in their territory. The development of a consistent waste policy was one of the first main issues to be handled by the regional governments. In this case study, we concentrate on the Flemish Region and their approach to the packaging waste problem. Two initiatives were taken in the Flemish Region in the beginning of the 90s with respect to this problem. One was the basic agreement on prevention, recovery, and recycling of packaging waste (26/6/1990). The other was the Flemish packaging waste convenant (26/3/1991). In this case study, we analyse these voluntary agreements to find out if they were effective and have led to results.

Environmental impact of the packaging waste problem
In 1991, 2,3400,000 tons of household waste were generated in the Flemish Region. This accounted for almost 423 kg/inhabitant/year. Almost a quarter (in weight) of that total amount was packaging waste. Thus, almost 500,000 tons of consumer packaging waste were generated every year in Flanders (beginning of the 90s). Furthermore, this amount represented almost half in volume of the annual amount of household waste. The industrial packaging wastes accounted for at least 1,000,000 tons each year. All kinds of materials were used in packaging

and could therefore be found in the packaging waste: paper, cardboard, metal, plastic (PVC, PET, HDPE), glass, etc. The normal way to dispose of packaging waste was landfill, sometimes after incineration. Less than 10% of all packaging waste was recovered as household waste. The methods of disposing of packaging waste at that time were generally considered a threat to the environment and a waste of raw materials and energy. Moreover, Belgium has a rather limited capacity for dumping. The high visibility of packaging in the overall amount of household waste, together with the damage these materials do to the environment, explains why the packaging waste problem has high priority on the policy agenda in the beginning of the 90s. The Flemish authorities have looked for a sustainable solution to that problem. They came up with a relatively new policy instrument: the negotiation of an agreement with industry concerning packaging waste.

Economic context of the packaging sector
This problem occurs in a very diversified sector. There are many different actors, and they often have diverging interests.

On the one hand, we have the *packaging producers,* which are often direct competitors. They produce equivalent packaging using different materials (glass, PVC, PET, metals, paper, cardboard, ...). And they all try to increase their market share to the detriment of their competitors. This strong competition is sometimes even encountered at an environmental level. Some products or materials have been touted as being better for the environment once they arrive at the waste phase. PVC, in contrast, is said to be very bad for the environment, although the scientific evidence is still disputed. The criticism of PVC as packaging material was mainly induced by some green parties who wanted to ban PVC as part of their global demand for a conversion of the chlorine chemical sector. This has led to the withdrawal of almost all PVC bottles from the market in Belgium. So we see that environmental considerations can have an impact on the competition between packaging producers and on their market shares.

The objective of these producers (to make a profit by producing as much packaging material as possible) is in fact completely contradictory to the objective of the voluntary agreements (to decrease the amount of packaging produced in order to decrease the amount of packaging waste). The association of packaging producers has always been preoccupied with the possibility that the government would intervene in the choice of particular packaging types for environmental reasons. Notwithstanding this contradiction, these producers have made great environmental efforts over the last ten years. They have introduced more environmentally friendly production processes, materials, and packaging, while making strides to decrease the volume of packaging. The involvement of some associations of packaging producers in the development of these voluntary agreements is a sign that they are willing to contribute to the achievement of the defined objectives.

On the other hand, we have the *fillers*. These companies buy containers and fill them with their products. These companies also operate in a very competitive market. They too often have divergent interests. In general, the fillers can be divided in two main groups. We have the multinationals (e.g., Coca Cola, Procter & Gamble, Unilever, ...) who produce the wellknown brands. They have a big market share and a rather strong influence on all levels of policymaking. And then we have the smaller companies (sometimes called SMEs) who produce products for the national or local market. They don't have as much influence on the development of policy.

This picture is not complete without the Belgian *distribution sector*. This sector includes wholesale and retail firms in the food and non-food sector. It consists of sole-proprietorships all the way up to large distribution chains. This sector makes a major contribution to the national wealth:
- 52 % of all consumer outlays are spent in the distribution sector;
- 10 % of the GDP is created in the distribution sector;
- 450,000 employees and businessmen are employed in the distribution sector.
This sector has always been very preoccupied with the threat of certain environmental policy measures in the packaging sector (for example, mandatory deposit return systems). They were always involved in the development of solutions for environmental problems at the regional, the national, and the international level.

Basic agreement on prevention, recovery, and recycling of packaging waste
This agreement, signed on June 26, 1990, was the result of negotiations between the Flemish authorities and different industrial partners. The authorities were represented by the Flemish Region (Minister of Environment) and the OVAM (Openbare Vlaamse Afvalstoffen Maatschappij -the public institution that is responsible for the planning and execution of the solid waste policy in the Flemish Region).

The industrial partners were VEV (Flemish Economic Association), NCMV (National Christian Tradespeople Association), the Professional Confederation of Sugar and By-products, the European Milk & Juice Carton Producers Association, FEDIS (Belgian Federation of Distribution Companies), LVN (Federation of Agricultural and Food Industries), Fabrimetal, FETRA, CLB, NVHDB, SIREV, the glass packaging sector, Association of Paper and Cardboard Producers, Fechiplast (Association of Plastic Processors) and BVI (Belgian Packaging Institute).

All signatories both industry and public authorities subscribed to the following objectives with respect to prevention, recovery, and recycling of packaging waste:

- a maximum prevention of the total amount of packaging waste;
- recycling as much packaging waste as possible;
- producing more environmentally friendly packaging.

This agreement was intended to foster a consistent approach to the packaging waste problem in the Flemish region, without promoting certain packaging types and/or certain products, by creating a co-operation between the Flemish authority and industry. With this agreement, both industry and authorities declared themselves to be openly in favour of a recycling strategy that gave priority to material and resource recycling (over energy recovery) as far as this is ecologically, hygienically, technically and economically possible. These objectives had to be realized by means of:
- the development of a Flemish Action Programme for prevention and recycling of packaging waste that included specific measures for prevention, recycling, and re-use of packaging waste, the establishment of a packaging database, and the initiation of a programme that is focused on household packaging waste;
- the foundation of a consortium to expand and finance this Action Programme;
- the negotiation of result agreements with the partners involved.
The industrial partners committed themselves to a series of initiatives. With respect to prevention of packaging waste, they promised:
- to look for and possibly apply environmentally friendly production processes;
- to reduce weight and/or volume of packaging (taking into account the product that is packed);
- to remove all substances that are harmful to the environment from the packaging;
- to intensify and if necessary initiative new scientific research in order to achieve these goals.
With respect to recycling of packaging waste, they promised:
- to include the recycling friendly aspect in the design, production, and use phase of packaging;
- to promote reusable, recyclable, and compostable packaging.
For both the authorities and industry, this framework agreement had to become the basis on which to enter into voluntary agreements with each other in implementation of the Action Programme. The partners preferred agreements to legal regulations as instruments to tackle this problem. That preference reflected the complexity of the packaging waste problem in the Flemish Region, the concern not to harm Flemish industry in international competition, and the fear of creating any distortions of competition or trade barriers. This agreement fits perfectly into the Flemish Waste Plan 1991-1995 that was focused on prevention, re-use, and recycling of all wastes in the Flemish Region.

Flemish packaging waste convenant - 26/3/91

The Flemish packaging waste convenant was an agreement between the Flemish authorities (represented by the Flemish Executive (Minister of Environment) and OVAM) and various industrial partners (VEV, NCMV, EMC Belgium, FEDIS, Fabrimetal, FETRA, Cobelpa, Fechiplast, SIREV, LVN, Cockerill Sambre, BVI, Federation of producers of non-ferrous metals, the glass packaging sector). This agreement is in fact the outcome of the implementation of the preceding basic agreement on prevention, recovery, and recycling of packaging waste (cf. supra). This convenant was signed on March 26, 1991 and remained in force until December 31, 1995 .

In implementation of the basic agreement, the non-profit organization PRO (Preventie- en Recyclingorganisatie voor Verpakkingsafval) was founded on June 19, 1991. It enjoyed mixed private and public funding (50 % public authority, 50 % industry). The goal of this organization was to become a consultative body for the packaging waste problem in Belgium and in that capacity to elaborate a prevention and recycling policy. This policy had to be focused on the prevention and recycling of packaging waste, the promotion of environmentally friendly packaging, the development of production and processing methods which are more environmentally friendly, and consciousness-raising among producers and consumers concerning the prevention of packaging waste and the recycling of the packaging materials.

In order to achieve this multifaceted goal, three different commissions had to be set up within PRO, each with a specific task:
- the Database Commission had to collect and provide data on packaging in Belgium;
- the Prevention Commission had to concentrate on the prevention of packaging waste in Belgium by elaborating research and by giving advice to the policymakers;
- the Strategy Commission had to concentrate on future developments in the packaging markets and packaging waste in order to work out a pro-active policy. The Action Programme of the basic agreement was further developed by this packaging convenant. The programme's objective was broken down into several topics with different goals:
- the development of a prevention policy for packaging waste, both on a quantitative (decrease the amount and volume of packaging) and a qualitative level (remove all hazardous substances from packaging). Furthermore, PRO had to develop measures to promote return systems on the administrative, financial, and physical level and measures to use clean technologies;
- the development of a recycling policy for packaging waste. This included the foundation of a sorting centre that could take care of the recyclable packaging waste of 500,000 inhabitants. The authorities would be responsible for the

separate collection and transport of these materials. The private sector would take care of the processing and the sale of the recycled materials. However, due to the 'Not In My BackYard' (NIMBY) syndrome, this sorting centre was never established;

- the promotion of the recycling friendly character of the packaging materials which are used;

- the promotion of the environmentally friendly character of the production and processing methods.

Main issues of the voluntary agreements on packaging waste
Motivation. The basic agreement and the Flemish packaging waste convenant were introduced because the packaging waste problem urgently needed a sustainable solution. Both authorities and industry have opted for a voluntary agreement, but each had their own reasons for choosing such a new policy instrument. The authorities took this route because of the complexity of the problem and the direct involvement of industry. (After all, industry is responsible for the creation of the packaging waste problem. With this voluntary agreement, they finally also became responsible for solving the problem.) Furthermore, it gives the authorities to a permanent voice in setting the environmental policy goals.

Industry, on the other hand, has chosen this option for a slightly different set of reasons. Like the authorities they recognized the complexity of the packaging waste problem in the Flemish Region. But the particular interests of the private sector are reflected in their concern not to harm Flemish industry in the international competition and not to create any distortions of competition or trade barriers. However, it is conceivable that the Flemish industry raised the argument of the protection of its competitiveness as a ploy to deter the authorities from setting more ambitious environmental policy goals. Under this voluntary agreement, the goals were not that ambitious. Industry had more time to adapt its environmental behaviour than it would have if a legal regulation had been issued.

Objectives and purpose. The objectives of both voluntary agreements remain rather nebulous. No quantitative objectives were included in the agreements, and no time frame for achieving the goals was set. Only some vague qualitative objectives were formulated (more prevention, more recycling, more environmentally friendly packaging). And these are very difficult to control and monitor. The obligations for both contracting parties remain rather limited, giving the final outcomes of the agreement little chance of success.

Implementation. These voluntary agreements were negotiated and implemented at the Flemish level. (Since the Belgian State Reform of 1980, the Flemish government is in charge of environmental policy in its region.) This limited area of application must be taken into account when the results of these voluntary agreements are evaluated. Moreover, Belgium (and also the Flemish Region) has a small open economy, completely integrated in the European Union, based on the free movement of goods and services. Due to its geographical situation in the centre of the Union, the country is also a hub of many distribution networks within the European Union. This central and open character of the Flemish economy surely has an impact on the results of the agreement. (In the Flemish Region, it's quite easy to escape from the obligations of the agreement by moving to the market of a neighbouring country (the Netherlands, Germany, France).)

Players / Stakeholders. As mentioned above, the packaging waste problem is situated in a very complex and diversified sector, with many actors who often have opposite interests. Therefore, it was decided to involve as many of these actors as possible in the development and implementation of the voluntary agreements. The industrial partners were represented by their associations or federations who signed the agreement on behalf of all their members. Working with associations or federations had two drawbacks:
- not all Flemish companies (dealing with packaging or packaging waste) belonged to a professional association or federation. The 'non-associated companies' were, for instance, not committed to these agreements and didn't had to take the objectives into account (the free-riders problem);
- not all the associated companies held the same opinion on the success and the implementation of this agreement. Some companies were convinced of the necessity of this agreement and its objectives and wanted this agreement to succeed. Other companies, however, didn't feel strongly committed to this agreement. They were not willing to make any effort with respect to prevention, recycling, and more environmentally friendly packaging (even though their association or federation had signed the voluntary agreement).

The many actors involved, the difference of opinion among associated companies about the success of the negotiated agreements, and the existence of non-associated, non-signatory companies had a direct impact on the results of the agreements.

Control and sanctions. Neither one of the voluntary agreements made much reference to control and sanctions in case of non-compliance. One reason might be the absence of quantitative objectives in both voluntary agreements. Some vague qualitative objectives were formulated, and these could hardly be monitored and controlled. There was for instance little incentive for the partners to meet the defined objectives. Both knew that non-compliance would hardly be punished.

Combination with other policy instruments. These two voluntary agreements on packaging waste were signed at the beginning of the 90s. Environmental activists were not very pleased with the outcome of these agreements. The voluntary agreements did not achieve the desired effects in the following years for the following reasons:
- the different motivation for authorities and industry to enter in this agreement;
- the vagueness of the objectives in the voluntary agreements;
- the complexity of the packaging waste problem;
- difference of opinion about these agreements among industry;
- limited solidarity regarding for this problem in the Flemish industry;
- limited availability of statistical data on packaging and packaging waste;
- complexity of the environmental legislation in the Flemish Region.
Therefore, these groups called for additional regulations in order to tackle the packaging waste problem in Belgium. This green pressure groups' request for additional regulations created a negative attitude towards the existing voluntary agreement. Because of this, industry lost all faith in the existing voluntary agreements and their implementation. That, in turn, made the results of the objectives even worse. Finally, this negative spiral of the voluntary agreements came to an end on July 16, 1993, when a system of ecotax was introduced in Belgium. The introduction of ecotaxes (on beverage containers and certain industrial packaging, for instance) meant the end of the voluntary agreements in Belgium. From that point on, industry was legally obliged to fulfil certain re-use and recycling objectives. Otherwise, they would have to pay ecotaxes. This federal (!) ecotax law undid all the efforts of the voluntary agreements and made the voluntary agreements and their objectives redundant. Therefore, the Flemish packaging waste convenant was never renewed.

Performance evaluation of the voluntary agreements on packaging waste
As mentioned above, no quantitative objectives were included in these voluntary agreements and no time frame for achieving the goals was set. Only some vague qualitative objectives were formulated. Therefore, it is very difficult to evaluate the effectiveness of these agreements. Nevertheless, we may assume that these voluntary agreements were not effective. After all, no clear goals were defined. From the very beginning, some of the companies involved expressed doubt that voluntary agreements would have an effect. Only two years after negotiating the packaging waste convenant, federal ecotaxes on packaging were introduced in Belgium (mainly because the results of the packaging convenant were minimal). These observations suggest that the voluntary agreements on packaging waste in Flanders were not effective at all. Also in this regard, no information is available on the cost/benefit of the targets set. Nor is any quantitative evaluation available on the cost effectiveness of the policy. In this case, even less can be said about

the agreements' efficiency. First of all, the policy objectives were not clearly defined in the voluntary agreements. Secondly, not much data were available on the implementation of these agreements. A more in-depth study would certainly be needed to evaluate the efficiency of these voluntary agreements on packaging waste. In both voluntary agreements on packaging waste in Flanders, no control body or follow-up commission was installed. Thus there was no one to monitor and control compliance with the stipulations of these voluntary agreements or the achievement of the goals defined therein. It was very easy for companies to ignore these convenants (free riders) because they weren't subject to control and no sanctions were defined in case of non-attainment.

6.5 Factors determining succes of negotiated agreements in Belgium

What conclusions can we draw from the preceding case studies? Although a more indepth study would certainly be needed, we may offer some tentative conclusions by testing the hypotheses, based upon a comparison of both cases. Apparently we have a rather successful agreement in the electricity sector and an unsuccessful one in the packaging sector. The characteristics of both cases allow us to advance an explanation for this striking difference.

a) the ESI and the public authorities have a tradition of negotiating with each other through many committees. That experience have made it easier to negotiate the agreement in a professional manner. In the packaging sector, on the contrary, there was no tradition of negotiating with the public authorities prior to signing these agreements.

b) Only a small number of negotiators were involved in the electricity convenant, namely the three Regions, Electrabel, and SPE. This considerably simplified the negotiation process. In both agreements on packaging waste, however, many different actors (often with diverse interests) were involved, which made the negotiations more difficult and time consuming. The electricity sector is dominated by one actor, Electrabel. This situation almost completely obviated the risk of free riders, which could jeopardize efforts to reach the goals in the negotiated agreement. The packaging sector, on the other hand, includes many actors (often competitors) with different interests. The free rider problem was thus a major thread to the realization of these packaging agreements. The openness of the Belgian economy only exacerbated this problem.

c) It is clear that the electricity convenant offered advantages to both parties (government and industry). The government, on the one hand, was enabled to achieve the goals it had set for this environmental problem. The results even went beyond the targets of existing legislation. Industry, on the other hand could pursue its long-term planning within a stable legal framework and could choose the most efficient way of dealing with emissions. In case of the packaging convenants, very few advantages were offered to the contracting parties. The government had very little assurance that this environmental problem would be solved by means of voluntary agreements (no quantitative objectives). And industry was not willing to make a major effort without some guarantee that their compliance would be rewarded (with no extra legislation) and that the free riders would be punished.

d) Realization of the targets of a voluntary agreement requires monitoring. The transparency and controllability of the electricity convenant is guaranteed by a follow-up commission which makes this convenant acceptable to third parties. The two agreements on packaging waste had very little transparency and controllability. No control body was installed to monitor and control compliance with the stipulations of both voluntary agreements to keep tabs on progress toward the achievement of the goals defined in these agreements.

e) It could be assumed that compliance with a voluntary agreement is more likely if there is a big stick behind the door. In other words, the parties are more likey to keep up their end of the bargain when there is a real and credible threat to the private sector that, in case of non attainment of the goals, the government will use other policy instruments. It is difficult to judge ex post whether such a credible threat existed at the moment of negotiation and during the life span of the agreement. Nevertheless, it is interesting that the electricity convenant explicitly states that the government can put an end to the convenant if the private sector does not fulfil the goals stated, at which point the government can use alternative regulations. In the packaging convenant, such an explicit statement was not made. However, we should note that in the packaging case, once the federal government unilaterally and unexpectedly took action by ecotaxes, the private sector could set up an impressive mechanism (FOST Plus) to recycle packaging in a massive way. Their initiative even anticipated a proper legal framework on the regional level. Apparently, they were hoping to be able to enter into negotiations at the federal policy level in order to obtain an exemption from ecotaxes.

f) One could assume that a voluntary agreement is more easily negotiated when there is consensus on the underlying goals and objectives that have to be reached. All parties concerned must more or less accept the reasons behind the policy targets and be able to reach agreement on precise reductions. Here too the dissimilarity between the electricity sector and the packaging sector is striking.

The environmental problem (acid rain) of the electricity sector dealt with in the electricity convenant is clearly defined. Overall consensus exists about the emissions which are responsible for it. Moreover, at that time, real targets were already set at the EU level and could be used as a benchmark. With respect to the packaging problem, many aspects were still hotly disputed at the regional level as well as at the EU level. For instance, the so called hierarchy of waste treatment was not yet firm. In particular, the preference given to re-use over recycling and to recycling over incineration with energy recovery was the subject of much debate. Other issues were the environmental merits or demerits of certain packaging materials, and the appropriateness of certain instruments (e.g,. the use of mandatory deposit refund systems). Needless to say, the more underlying conflicts of ideas there are, the more difficult it is for an agreement to be reached.

7. PARTNERSHIP AS A LEARNING PROCESS

Environmental convenants in the Netherlands

PIETER GLASBERGEN

7.1 Introduction

Dutch environmental policy of the past decade may be seen as a quest for new relationships between government and the private sector. What makes this quest so special is the clarity of its mission: to make the problem-solving capacity of government more effective when dealing with environmental issues. Yet the aim is not to expand the reach of government agencies. Nor does the policy seek to saddle the private sector with responsibilities that were previously public-sector tasks. Rather, it is an effort to give established relationships between public and private domains a new impetus. On the basis of existing responsibilities, which are recognized as being different, the relationships are to play a productive role in tackling environmental issues.

To realise this aim the environmental covenant is introduced as a new policy instrument.[1] The essence of an environmental covenant is that at least two parties -of which one is a government agency and the other a representative of a sector of industry- reach an agreement on the realization of environmental targets. In some cases, other civil organizations such as trade unions or environmental organizations may also be involved. Having consulted with each other the participants in the deliberations put down in writing their arrangements on how to tackle the environmental problematic. Particularly novel is that the agreements pertain to future actions, to be carried out by the parties, in mutual dependence on each other. The future actions are aimed at jointly implementing the environmental policy initiated by the government.

The Netherlands was one of the first countries to carve out an important niche for environmental covenants. By now, the number of formal agreements that have been arranged is well over 100. Meanwhile an environmental covenant has become much more than an arbitrary attempt to try out a new policy instrument. Although it did start out as a mere test case, the application of this instrument has

---

[1] This instrument is also known by various other names: a contract, an accord, a statement of intent, and a policy agreement. The name is less important than the function, of course.

P. Glasbergen (ed.), Co-operative Environmental Governance, 133–156.
© 1998 *Kluwer Academic Publishers. Printed in the Netherlands.*

gradually led to a new concept in governance. An essential characteristic of that concept is the effort to develop a policy that connects the assumed collective advantages of environmental policy with the individual (economic) possibilities and limitations of companies and sectors of industry.

In this chapter we analyse the development of this new concept of governance as a learning process. 'Learning' refers to the changes in the application of covenants. While working with covenants the main actors learned when and how to deal effectively with the instrument. This learning process consists of three phases, which overlap in part. After a discussion of these phases, our analysis turns to a critical evaluation of governance by covenants. It is stressed that the environmental covenant is no cure-all, but, under certain prerequisites, fulfils some positive functions other policy instruments cannot fulfil.

7.2 Covenants as gentlemen's agreements

The first environmental covenants were introduced in the second half of the 1980s. These were single-issue and for the most part single-actor covenants. They were the vehicle by which the Ministry of the Environment made agreements with a company or a body representing a type of industry regarding a particular polluting substance or a certain product. Examples are rife: the covenant signed with manufacturers on the production of detergents without phosphates (1987); the covenant signed with soda bottlers on charging a deposit on PET containers (1987); with the Netherlands Aerosol Association on the use of CFCs in spray canisters (1988); and with bottlers of beer and non-alcoholic beverages on the use of cadmium in crates (1988). By giving a concise description of one case (see also Klok, 1989), we can amply illustrate this phase.

Phosphates in detergents
At the end of the 1970s, the government held preliminary discussions with the Netherlands Association of Soap Manufacturers (NVZ) regarding the substitution of phosphates in detergents. The manufacturers indicated their willingness to cut back the proportion of phosphates in their products. The goal was to reduce the share by 40 percent between 1980 and 1983. In view of a number of new technical developments, it was expected that this target could be reached without much difficulty. And indeed, the reduction was made. By the mid-1980s, this environmental issue reappeared. By then, it was a more politically charged point on the agenda. Protest actions by environmental and consumer organizations got broad coverage in the media. They pointed out the Swiss legislation (dating from January 1986) prohibiting detergents containing phosphates. Similar legislation was called for in the Netherlands. The protest groups were supported in their

efforts by the newly installed Minister of the Environment. In his opinion, it was about time to take more far-reaching steps, and he made that standpoint public. Subsequently, both parties consulted with the NVZ with the aim of setting a concrete time frame for the replacement of phosphates in detergents. At that time the producers were reticent to take any action. From both a technical and an economic perspective, they considered substitution to be feasible only in the long run. At that point, facing the threat of an embargo and heightened public protest, the manufacturers felt the pressure building up to make concrete agreements. The increasing pressure was not what led them to concede, however. The turning point came in 1987, when one of the big soap manufacturers put a name-brand phosphate-free detergent on the market. A number of supermarket chains followed suit with their own house brands. The same year -actually, in just a few months- an agreement could be reached. In this covenant, the NVZ committed itself to make an effort to increase the market share of phosphate-free detergents by 40 to 50 percent within a year. The commitment was contingent upon analogous development in other countries. After 1989, the parties would promote the continuation of the process of phosphate substitution in detergents. In order to stimulate the public to buy detergents without phosphates, various promotional measures were taken. The government was going to monitor the development in the product market share and take appropriate action if necessary. But the need did not arise. Soon after signing the covenant, the development of the consumer trend toward using phosphate-free detergents picked up momentum. As it turned out, there was demand for the product. The commitment to make an effort proved to tie in well with the commercial interests of the manufacturers.

This case clearly reveals the main characteristics of the initial environmental covenants. In this phase, the covenants tend to serve a strong symbolic function, the policy is usually ad hoc, and the impetus of policy is limited.

Symbolic function. The motivation to sign a covenant is generally provided by environmental and consumer organizations in the form of a protest campaign. Government policy regarding the environmental issue in question is already being developed. But the demonstrations are the main reason why the topic is placed high on the political agenda. The relevant minister picks up on the issue and sees an opportunity to get a higher profile in the public eye by signing a covenant. Political representatives cooperate with environmental and consumer groups, even though these civil organizations are not formally party to the covenant. The main thrust of that cooperation is to demonstrate a willingness to tackle environmental issues forcefully.

Ad hoc policy. In connection with the previous point, it is clear that the political approach to the issues is arbitrary. There is no policy-driven strategy to support decision-making with reference to when it might be expedient to arrange a covenant. Regarding the polluting substance that prompts the covenant, other applications are often not recognized. From an environmental perspective, it might be more critical to take completely different products off the market. Yet in those cases, no covenant is made. It is noteworthy that in those covenants that are actually signed, the relation to the wider policy to deal with the pollution is hardly touched upon, if mentioned at all. In the background, however, there is always the possibility to achieve the aims of the covenant by introducing general regulations.

Limited policy impetus. The scope of the covenant is already limited. Yet it can hardly be said that the policy impetus is strong. The covenant represents more or less the final stage in a policy process that has already run its course. This instrument formally confirms the options that are both technically and economically feasible. Various alternatives are available, and the companies involved are already moving in the desired direction. They are motivated to do so by market demand. It is noteworthy that other countries are sometimes further along with respect to the contents of the covenant. Before the Dutch covenants were arranged, the Swiss had already prohibited the use of phosphates in detergents. And in Germany, to give another example, an agreement was already in place to terminate the use of CFCs in aerosol cans.

7.3 Toward a wider scope for covenants

In this phase, the Netherlands is a leader on the international stage with regard to the introduction of agreements in environmental policy. But this is not necessarily a progressive stance with regard to the content of the policy or its compelling nature. The covenants are ad hoc solutions to problems that occur ad hoc (Biekart, 1993). Even without a covenant, the envisioned environmental aims would have been achieved. Moreover, it is more accurate to characterize the first environmental covenants as gentlemen's agreements than as real contracts. In a legal sense, their status is ambivalent; for instance, the legal construction and the enforceability have not been tested. These topics set the stage for discussion. In the next phase of application of environmental covenants, in fact, they are the central issues. Be that as it may, the idea has taken root that the covenant is a vehicle by which policy can be brought to bear in a quick and effective manner. The covenant also conforms to a new philosophy of governance that is gradually taking shape. The key word in that development is 'internalization'. This term denotes that if the process of environmental change is to continue, the private sector itself will have to tackle the problems. This standpoint corresponds with agreements between the government and an accountable private sector. By accepting responsibility, the business community also has the opportunity to put its other interests at stake. In an ideological sense, the parties can communicate because all participants share the opinion that a process of environmental change must go hand in hand with an effort to maintain a competitive industry.

Within this framework, an attitude is cultivated in which a fruitful quest for cooperative relationships can take place. This is manifest in the signing of dozens of new environmental covenants, by various ministries, in the course of a few years. Representative bodies at the provincial and municipal level are increasingly drawn into the process too and are often co-signatories of a covenant. The character of the covenants is also changing. Single-issue and single-actor covenants are still arranged. One example is the covenant on emissions of substances that contribute to acid rain (SO_2, NO_x), which was arranged with the power companies (1990). Another example concerns soil sanitation at gas stations; the oil companies were party to this covenant (1991). Yet another example is the covenant on the discharge of amalgam residue by dentists (1991). In addition, a second category of covenants is emerging. Although they are still single-issue agreements, they cover more complex branches of industry. An early example is the series of agreements that were initiated by the Ministry of the Environment in the mid-1980s. Agreement was reached with ten or so branches of industry to reduce the output of hydrocarbons (HC) and volatile organic compounds (VOC) (see below). Another example is the packaging covenant (1991), which envisioned a ten-percent reduction in waste material over a ten-year period. The party to this covenant represents 150 firms.

HC 2000

Hydrocarbons have been implicated in the depletion of the ozone layer. The trouble is that they are emitted from many sources. To make matters worse, at the end of the 1980s, it was not known how to reduce emissions. Under such circumstances, unilateral regulation is ineffectual. This is why diverse government organizations met with representatives of the private sector to hammer out a strategy to reduce emissions. The reduction was supposed to be accomplished step-by-step in the period 1988-2000. The target that was agreed upon was to bring annual emissions of volatile organic compounds (VOCs) down from 225 to 100 kiloton. Representatives of government and industry translated the target into an implementation strategy. A national steering committee was set up to direct the whole process. And for each relevant branch of industry, a task force was appointed to guide the participants along the way. To support the implementation, dozens of studies were carried out; development projects and demonstration projects were set up; conferences, courses, and workshops were held; brochures, reports, and fact sheets were written, and so forth. Public-private cooperation is of interest to the government because it entails an active input of knowledge on the part of industry. And cooperation is of interest to the business community because it keeps them informed on where they stand with regard to government policy. At the same time, they get a reasonable amount of time to work out the measures themselves. In the course of time, a steady reduction was achieved: 11 percent in 1990; 22 percent in 1992; and 25 percent in 1993. Meanwhile, industry gained deeper insight in the actual amount of emissions. They realized that greater reductions would be necessary to be able to meet the final target in the year 2000. A number of industrial sectors booked good results, others showed less progress. Specifically, the aluminum branch could not keep up with the other sectors because this sector is subject to stiff competitive pricing imposed by the international markets. In the printing and paper industry, the dependent position between suppliers and customers plays a crucial role. In the paint and pigments industry, market demand plays an important role (Van Vliet, 1993, pp. 112-116; Van de Peppel, 1995, Chap. 8.)

The application of the environmental covenant -which might be described as exuberant- still takes place for the most part without a clear policy strategy. There are major differences in form, function, and effectiveness. Some covenants are no more than non-committal agreements to study the issue. Others are a statement of intent to reach certain environment goals. And yet others contain agreements that might be called obligations to achieve given results. There are also major differences with regard to regulations and stipulations concerning topics such as implementation monitoring, conflict resolution, options for change and termination, evaluation, and sanctions on non-compliance. With regard to the involvement of the parliament, major differences also exist. Only in the course of applying the covenants does it become clear what the foundations are for a policy in which the environmental covenant will take a more or less fixed place. Two key issues are still pertinent:
a) the position of the environmental covenant in the legal system;
b) the efficacy of environmental covenants;
The following sections described these two issues in some detail.

7.4 Position of the covenant in the legal system

Ever since the first environmental covenants were signed, one question has been awaiting an answer. How is the covenant, as an agreement in private law, related to regulations under public law, such as environmental legislation? This question is particularly important in situations whereby different levels of government perform different tasks and have different responsibilities. In the Dutch system of environmental law, the ultimate responsibility for the implementation of the national environmental policy lies with the local and the provincial authorities.

These government bodies are supposed to translate the national environmental aims into permits for individual companies.

A permit is an instrument that can be imposed unilaterally by government decree. The permitting process has fixed procedures. And it is tied to opportunities for third parties (unrestricted) to lodge objections and appeals. The assessment of content -that is, the nature of the check to see if production activities comply with environmental requirements- also has to follow formal procedures. The environmental legislation requires application of the ALARA principle. This means that companies are expected to apply those techniques whereby the environmental load will be as low as reasonably achievable. Like the related and much used concept 'state of technology', the ALARA principle provides a standard for what may be considered reasonable. It should be possible to expect a company to take steps in light of the fact that a given measure has already been tested elsewhere and has been shown to be financially acceptable. There is some scope for negotiation, though narrow.

In contrast to a permit, a covenant is a bilateral instrument. It is based on building consensus in formal and unlimited negotiations. There is no standard procedure to arrive at a covenant. Nor is there a formal option for third parties to exert influence. Furthermore, it is important to note that the parties involved in a covenant -the national government and the branch organizations-are not the same ones as those that are involved in the procedure to grant a permit to a company.

In order to determine the position of a covenant in the legal system, two situations must be taken into account. The first is that in formulating a covenant, a government may deviate from the formal requirements stipulated in a permit that has already been granted. The second situation is that a covenant might be arranged at one level of government whereas a lower level of government may make different requirements. Environmental organizations have gone to court over the first situation. Their criticism revolves around the use of a covenant as a means to relax environmental regulations previously formally ratified. One of the most typical cases is the following.

Kemira Pernis (1991)
This case refers to a firm that produces artificial fertilizer. On the grounds of the Pollution of Surface Waters Act (WVO), this company needed permission to discharge effluent. In view of the location of the discharge -a national body of water- the Minister of Transport, Public Works and Water Management is the designed authority to issue a permit. A covenant that had been arranged between this company and the ministry specified a gradual reduction in the cadmium content. On condition that this agreement would be met, permission was granted

to discharge effluent. Nevertheless, an environmental organization took the parties to the Administrative Court and won the repeal of this permit. The verdict was that according to current legislation, the request for permission to discharge would have to be reviewed with regard to upholding standards for water quality. An audit along these lines would have to be based on available insights in environmental hygiene. Agreements couched in a covenant do not fit into the legal system. That system does not leave any leeway for circumventing the required assessment on the grounds of environmental hygiene (Van den Broek, 1993).

The second situation that is pertinent here is one in which a government agency has arranged a covenant with a body that represents a category of companies and, at a later date, the individual companies are confronted with different and more stringent requirements. This situation has arisen on several occasions. Specifically, it occurred with regard to the elaboration of the HC 2000 agreements, described above. In this case, it was the companies that called for legal proceedings. They felt that expectations had been raised by central government that were not being lived up to.

HC 2000
Earlier in this chapter, we referred to the general aim of the HC 2000 project: by the year 2000, emissions of volatile organic compounds should be half the level of 1981. This aim has been worked out for each industrial sector in the form of a specific target and a relevant package of measures. One question arose in the process: Could individual establishments invoke the covenant? In particular, establishments that had to meet specific requirements at the time permission was granted have gone to court over this. For instance, one firm was required to reduce emissions of organic solvents by 90 percent. In the covenant for the branch, a 75-percent reduction was considered sufficient. Another enterprise lodged an appeal against specific requirements. They were of the opinion that it was not fair to determine compliance with the desired reduction targets at the level of the individual enterprise. Instead, they argued, the assessment should take place at the level of the industrial branch, summing up the emission reductions achieved by individual firms. A third firm lodged an appeal against the requirement imposed by the municipality to develop an implementation plan for the enterprise on the basis of the best practicable technology. That plan would have to be carried out within a specified time frame. From the point of view of the firm, these demands would contradict the outcome of the national consultation on implementation of the covenant. The Administrative Court threw out all of these appeals. Their reasoning was that the government agency in charge of granting the permits has its own jurisdiction grounded in the Environmental Protection Act. The verdict reiterated the standpoint that a covenant between the national

government and the private sector cannot contravene that authority (Van den Broek, 1993).

In both of the situations reviewed above, we see that the introduction of the covenant has called the reliability of the government into question. The environmental movement was concerned about a relaxed enforcement of environmental policy that in many cases had been the outcome of political strife. The private sector was concerned about the reliability of the government as a partner in consultation and negotiations. The judgements in the court cases have clarified the position of the environmental covenant as an instrument of environmental policy. Basically, a covenant is a voluntary agreement that cannot contradict the system of public law. In that light, there is no free choice between an agreement under private law and the route of public law. It is possible to use a covenant to anticipate upon regulations that have not yet been formulated. And it is also possible to use a covenant to supplement existing legislation. But a covenant can never replace something that has already been established in public law.

7.5 The efficacy of a covenant

The second question refers to the efficacy, within the limits outlined above, of that which is regulated by a covenant. Legislation is predicated upon on a system wherein the quality of proposals that have been formulated on the grounds of detailed instructions have to be tested several times. For contracts under private law as well, from a legal angle, there are more or less fixed requirements. In a sense, covenants may also be considered contracts. Yet the situation is somewhat more complicated than for ordinary contracts. In covenants there is often a mix of statement of intents to make an effort and to achieve specified results. As pointed out earlier, there is great diversity in the way a covenant is worked out in terms of the conditions that are necessary to meet (monitoring, evaluation, sanctioning, etc.). It remains unclear whether or not a covenant creates contractual obligations that will hold up in a court of law. If the legal enforceability is not guaranteed, the efficacy of the instrument may be dubious. The efficacy of a covenant is bounded by two factors: its voluntary nature and the authority of the parties. Each of these points warrants further discussion.

Voluntarism
A covenant is arranged between parties who perceive one another as equals. They may be seen as partners. Although the threat of a more coercive stance on the part of the government may play a role in the background, the partners enter into

the arrangement of their own volition. Within the realm of voluntarism, the interests of the parties will play a crucial role. A covenant will only be signed if each of the parties involved can serve their own particular interests by doing so. In brief, the benefits of a covenant must be reciprocal.

The interest of the government is primarily to pursue its environmental aims. In the context of these aims, the government employs the covenant as an instrument of policy. In that respect, the politically sanctioned policy aims serve as a standard on which to assess the value of the agreement. This constitutes a significant limitation of the meaning of the policy instrument. Specifically, a covenant must offer some prospect of a change in behaviour.

For the private parties, the policy aims are in principle only meaningful to the extent that these aims affect the actors' own interests. The certainty they seek is of a different nature. It is mainly determined by financial and economic interests.

In the course of consultation and negotiation over possible covenants, there will thus always be some degree of tension between a government that wants to achieve policy aims and a private party who has other interests to defend. Besides serving one's own interests, the private party often has an interest in weakening or delaying progress on the environmental aims. The result will be a delicate balance. Each party will have to judge whether sufficient certainty has been gained in light of one's own interests. That certainty will be less important if the government policy is still not very explicit and/or uncertain policy situations exist.

A covenant can also be functional under such conditions. However, the agreements would have to be very general in nature and perhaps even ambivalent in its legal wording. The overall character of the covenant will then be an agreement to get together and work out a jointly determined development trajectory within a set time frame.

Unlike legislation, which in principle must be enforceable with no loopholes -on paper, at least- a covenant derives its value in part from the opportunity to circumvent that requirement. The flexibility this creates for applications of the instrument is precisely its value. An assessment that is based on legal technicalities has only limited significance when flexibility is required.

Authority of the partners
A second limitation of the efficacy of a covenant lies in the characteristics of the parties involved. The scope to make policy is formally restricted for governments and private parties alike. It was pointed out earlier that a contracting minister cannot make any agreements that lie under the purview of other government bodies such as the province or the municipality. However, there is a more profound issue at stake. Government competence is always democratically controlled authority. Discussions on the main issues of policy should be conducted publicly in parliament. The consultation should not be held in closed circles and

only reach the ears of parties with a direct interest. Closed consultation is always an option and often serves a function. But it has to lead to formal political decision-making in a public arena. The minister who arranges a covenant is always restricted by the authority of the parliament. Even when there is political consensus at the time the covenant is signed, later changes in political insight may be considered to weigh so heavily that unilateral measures may still be taken. Other government authorities who are party to a covenant may also have limited scope for policy-making. The representative bodies at the level of the province and the municipality have no hierarchical authority with regard to the organizations they represent. For them too, the covenant is not unconditionally binding, unless each province and every municipality is co-signatory to the agreement. As yet, this is not the case. With regard to the municipalities, it would not be practical either, because there are so many of them.

The private party also has limited room to maneuver with regard to policy-making. A private party who represents a category of enterprises or a branch of industry usually has no hierarchical authority with respect to the group represented. Branch organizations are generally associations that embrace all the enterprises. The authority to sign contracts in the name of those enterprises is limited. Only when individual enterprises are co-signers -as in some recent covenants (see below)- is there greater certainty about whether or not the parties can be held to the agreement.

Thus, the extent to which a covenant is binding for the parties who sign it is only limited. At the same time, it should be noted that certainty in this respect is often still limited even when the regulation is by all appearances watertight. A telling example is the covenant on tropical wood.

Tropical wood
This example concerns a covenant with diverse signatories. Four ministries (the Environment; Agriculture, Nature Management, and Fisheries; Economic Affairs; and Overseas Development), the lumber industry, two nature conservation organizations, and the trade unions signed the covenant in 1991. The aim was to restrict the trade and processing of tropical wood in the Netherlands. From 31 December 1995 on, the covenant only granted licenses to process wood that came from countries or regions with a forestry policy geared to conservation of forests and sustainable production of lumber. The parties indicated that the covenant had to be seen as an agreement within the framework of civil law. One of the main activities was to introduce a certification system (seal of authenticity). The other was to liquidate stocks of lumber that had not been produced by sustainable methods; the stockpiles were to be eliminated by the time the covenant was to have run its course. This covenant is a classic example of a technically sound agreement. Its targets are clear, the time frame is specific, compliance is

monitored, evaluation takes place at fixed points in time, and there are rules for opting out of the agreement as well as for settling disputes. The covenant even stipulates which sanctions would come into play if the parties did not keep up their end of the bargain. Enterprises that break the agreements will be fined if the stocks have not been liquidated by the time the covenant comes to an end.

Notwithstanding this sound construction, implementation of this covenant turned into a complete fiasco. Various reasons can be identified. The main problem was ambiguity about what sustainable forestry policy actually is. An advisory committee that was appointed to define it could not determine how to measure sustainability. Furthermore, the nature conservation organizations proved unable to muster support for their proposal to make do with a working definition for the time being. Then, they promptly withdrew from the covenant. Besides uncertainty, another factor also undermined the covenant, namely a conflict of interests among the ministries. Most importantly, the Ministry of Economic Affairs opposed all far-reaching measures, such as an import embargo on lumber that was not produced by sustainable methods. The ministry was supported in this stance by some wood-exporting countries, namely Malaysia and Indonesia. These wood-producers were strongly opposed to the covenant. To complicate matters, the coalitions of ministries that supported each other keep shifting. The Ministry of Overseas Development, for instance, was initially in favor of an import embargo but then turned against it. Furthermore, the evaluations ran into delays when the ministries could not agree with each other. The parties now console themselves with the thought that they have gained insight into the complexity of the issues. That insight may prove useful in formulating international rules. And that regulation is currently in preparation. It is thus only a matter of time before certification will be introduced.

7.6 Toward a profile for environmental covenants

Covenants are made on a voluntary basis and with limited possibilities for commitment by the parties. This explains why the accent is on the statement of intent. The legal criticism is mainly concerned with this aspect of the covenant. Formally, the legal grounds of this instrument are shaky. The question is whether this instrument is weak in practice. It has gradually become clear that the power of this instrument lies in the specific function that it can perform within the realm of environmental policy. From the government's point of view, that function lies mainly in the legitimation of government policy and the support it musters for carrying out the environmental aims. For the private sector, the function lies mainly in the opportunity to draw in a company's own interests, which are mostly economic. Government policy can be formulated in such a way that private interests are also secured. Both parties can use the covenant to create some degree of stability in their future relations. They cover themselves against arbitrary actions by parties on whom they have become highly dependent. The fact that they do so voluntarily, with only a limited legal commitment, is an inherent characteristic of the instrument. From the perspective of the government, the point is to foster a change in conduct. It is not the aim of the government to force that change, as legislation purports to do (Aalders, 1993). In this way, the parties can commit themselves as concretely as they themselves desire, to the extent they are able to do so and perceive to be useful. Even with these restrictions, however, the agreements still have some value. Voluntarism implies a self-imposed engagement. Much more important than the firmness of the legal basis is the moral commitment. Without a commitment, any more formal security has no meaning (Konijnenbelt, 1992).

From this perspective, it is important to first estimate the reliability of the partners. If it is possible to make agreements with reliable partners about sharing responsibility for the implementation of the aims of environmental policy, then in principle the chance of achieving the targets is greater than it would be when environmental measures are unilaterally imposed. Secondly, it is important that the parties are clear about their intentions. Here too, legal requirements on the form are less important than a code of conduct that marks the issues that need to be arranged. One of the aspects of a covenant that makes it work is its flexibility. Any legally precise formulation would undermine the utility of the environmental covenant. Against this backdrop, the Dutch government decided not to provide a legal definition of the covenant as an instrument of environmental legislation. Instead, a memorandum was prepared giving instructions and identifying issues that warrant attention.

At the outset of this chapter, we characterized a covenant as an agreement between a public party and a private party that is intended as a means to achieve government policy aims. As an intermediate step in our analysis, we can now describe in greater detail the profile of an environmental covenant as an instrument of policy, in the definitive form it has gradually taken.

- First of all, the covenant is an instrument that has no formal place in environmental law. There are, however, standpoints about the scope of its application that have been made more explicit. According to those standpoints, a covenant can in principle only be applied if doing so may be expected to lead to a greater effectiveness than the application of other instruments. Formal unilateral regulation forms the basis of environmental policy. The covenant performs a role as supplement to and articulation of legal frameworks.

- Secondly, the covenant is an instrument that presumes an environmental issue that can be clearly circumscribed or an environmental policy target that is clearly specified. With regard to this issue or target, it should be possible to distinguish well-organized parties. Representatives of those parties are supposed to represent their constituencies. The covenant can thus only be applied in an institutionalized field of actors in a social arena.

- Thirdly, the covenant is an instrument that is based on voluntarism. Its power is based more on social control than on legal regulation. It operates through rules of conduct and business codes. Regarding the technicalities of a covenant, the most important aspect is that the parties must be explicit about what they want and must be clear in their agreements. Commitment can be fostered by imposing definite time limits, a monitoring system, and an assessment system. These are fixed elements of every covenant. Other stipulations may be important too, such as rules on arbitration and sanctions. However, in view of the nature of the instrument, the enforceability of a covenant in a court of law is less significant.

- Fourth, the covenant is an instrument with a reciprocal character. This implies that the consultation and the negotiations between the parties are aimed at making productive use of their mutual dependencies. The parties seek to establish what might be called a strategic alliance (Hoekema, 1994; Bressers and Klok, 1996). This is only feasible if antagonistic relations are transformed into relations that promote cooperative action. A covenant is thus a consensus-seeking instrument that is deployed to make the implementation of policy more viable.

7.7 The environmental covenant in target-group policy

At the beginning of the 1990s, the position and the role of the environmental covenant changed again. The multiple-issue and the composite-actor covenant came to the fore. The main factor triggering this shift was the introduction of the first national policy plans. There was a definite trend toward a cleaner society (Nelissen, 1990). Yet from the perspective of sustainable development, there was still a need for major policy efforts to bring emissions of polluting substances down to an acceptable level. This effort was not expected to clean up individual and incidental cases of pollution. Rather, it envisioned a different emission profile for industrial sectors as a whole. For most contaminants, a reduction in emissions by 70 to 90 percent was considered necessary. According to the national environmental policy plans, these levels must be reached by the year 2010.

With regard to governance, the question now became how to motivate large social subsystems to make such a drastic long-term change. Large industrial sectors have their own dynamic and development trajectory. How can that be dealt with while doing justice to the interests of the sector and yet also pursuing ecological renewal? This governance question is fundamentally different than before. The instrument of environmental law seems hardly suitable to satisfy the new need for governance. Formally, each government that is charged with the task of granting permits should have to look for possibilities in each individual establishment to implement the policy step-by-step. Apart from the practical possibilities, the sheer number of parties involved increases the chance of inconsistency in implementation. Therefore, the outcomes of the licensing track have to be classified as highly unpredictable. The uncertainty is especially due to the need to take the financial and economic possibilities of the enterprises and their industrial branches into account. Against this backdrop, a new policy strategy is being sought, one in which the covenant plays a key role. That strategy is an elaboration of the policy philosophy previously developed around the concept of internalization. That notion expresses the desire to stimulate the policy objects themselves to take responsibility for tackling the environmental issues. The target groups of the policy are supposed to be actively engaged in finding a solution to the problem. The role of the government can then be directed to the structuring of the process of change, to the creation of stimulating conditions, and, of course, to the adjustment of the process, if the need should arise.

A new policy strategy
The experience with HC 2000 was fairly encouraging. Therefore, the approach used in that context is now being worked out for the industry as a whole. At that level, the approach is called target-group policy. For diverse branches of industry, government agencies and branche organizations representing a sector of industry are making a concerted effort to develop an appropriate structure for implementing environmental policy. It should allow a stepwise translation of the national targets for emission reduction to targets for sectors of industry. In a covenant, the parties formalize this structure in a number of central elements:
- an Integral Environmental Target Plan (IETP) for the entire branch of industry, stipulating the final targets to be achieved and the interim targets to be reached along the way;
- an Implementation Plan that specifies the IETP for the entire branch of industry, or a Company Environmental Plan (CEP) that does so for each individual company;
- an annual report for each company about the progress along the emission reduction trajectory;
- a Branch Consultation (BC), held among the relevant government officials and representatives of the branch of industry, that guides the entire process.

The process of change that is envisioned to come out of these elements is incremental in character. It is explicitly intended to be cyclical in nature. Essentially, the aim is to set a continuous learning process in motion.

Step 1: setting the targets for the industrial sector
The first step involves translating national policy aims into industry-wide targets in general and specific branches of industry in particular. In a structured consultation among government bodies (ministries, municipalities, and provinces) and representatives of the branch organization, a statement of intent is formulated. That document includes an Integral Environmental Target Plan (IETP). The term 'integral' implies that the aim is to achieve total results in the long run in the reduction of pollution of air, water, and soil and to make progress in the areas of energy-saving, waste materials, soil sanitation, and so forth. Besides the ultimate goals, specified in terms of reduction percentages relative to a baseline year, there are also interim goals (1995-2000). In principle, the national policy goals are not open to negotiation. It is possible, however, for the branch to offer constructive criticism on the technical and financial feasibility of these goals. Moreover, the time frame (with 2010 as the fixed end point) is open to discussion. When agreement has been reached on the IETP, it is signed by the relevant government officials, the representatives of the branch of industry, and -on a voluntary basis- now also by the individual companies (within heterogeneous braches, see below). In this manner, the parties make a limited commitment; they sign a statement of intent. The only obligation to achieve results refers to the subsequent steps that are to be taken.

Step 2: operationalization on the company level
In the second step, the IETP is then worked out at the company level. In the process, a distinction is made between homogeneous branches of industry and heterogeneous ones. Homogeneous branches include companies with a limited number of similar industrial processes (such as the paper and printing industry). This makes it possible to take a standard approach to the environmental problems. For this category, an implementation plan is formulated at the level of the industrial branch. An implementation plan is, in fact, a handbook defining packages of measures and giving standard rules for permits. The contents of such plans can be directly implemented by the competent government authority - usually the municipality- in the trajectory of subjecting permits to review. For the heterogeneous branches of industry, this standard approach is not feasible. That is because these branches consist of large and complex companies that are usually confronted with a multitude of environmental problems. The options for resolving environmental problems differ widely from one company to the next (as in the chemical industry). Here, it has been decided to deal with each company on an individual basis. The covenant includes the obligation to translate the IETP into a Company Environmental Plan (CEP). In the CEP, the companies will include those measures that are in accordance with the state of technology. In selecting which measures to include, however, priorities may be set on the grounds of

financial and economic considerations. Thus, the CEP is not necessarily a one-to-one translation of the IETP. The basic assumption remains that all final goals will be achieved by the year 2010. The company itself is supposed to show initiative and develop a plan. The CEP is formulated in close consultation with the government body that is in charge of granting permits; in most cases, this is the provincial authority. Furthermore, the CEP is subject to approval by that same authority. The plan is in force for a period of four years. In fact, the CEP may be considered to be a derived covenant. At the same time, the companies commit themselves to submit an annual report to the competent authority describing the progress that has been made in dealing with the problems. Moreover, targets can be included in new permits or in permits that come up for revision.

Step 3: feedback and monitoring
The third step involves feedback of the results of the CEPs to the IETP. By totaling the results, it is determined whether the targets specified in the IETP can be reached on schedule. To perform that task, a special consultative body is instituted, the Industrial Branch Council. This is a joint consultative body consisting of government officials and representatives of the branch of industry in question. The council is charged with the responsibility of guiding the entire process, monitoring the implementation of the IETP, verifying the results, conducting studies on bottlenecks, and looking for ways to resolve the obstructions. This consultative body is not concerned with individual companies. Rather, it is explicitly concerned with the target plans of the branch of industry as a whole. In principle, the task of finding bottlenecks and figuring out how to get rid of them is intended as a stimulus for the companies that belong to the branch. The recommendations that have to be made are partly directed to the next round of CEPs. In this way, the cyclical character of the process becomes evident. The final goals can be attained step by step, in a learning process. The consequence is, however, that when the CEPs are formulated the first time, the IETP might only have a limited impact. If bottlenecks emerge in the implementation of the IETP goals, then it might be preferable to work up to a more stringent application of the policy in a subsequent round. The consultative body submits a report on its activities to the Minister of the Environment, among others.

By setting up this broad framework, the foundations have been laid -at least in principle- for a fruitful collaboration between government and the private sector. The core elements have been defined. The activities are tied to deadlines. In addition, the results are implicitly public; the plans and progress reports are available upon request. With the target-group approach, the environmental covenant has taken on a new function in policy. As an independent policy instrument, the covenant has a less central position. The closure of a covenant has become a formal step in a broader policy strategy. The guided process of change is more important, and the covenant is embedded in that process. The covenant specifies the structural and processual elements that are crucial to making progress.

7.8 Effectiveness of covenants in the target group policy

With target group policy a more collaborative government and a more collaborative branch of industry have arranged their reciprocal relations. In this way they can make their mutual dependencies productive. The approach has been worked out first for a number of key branches -also the biggest industrial polluters- of industry. These are basic metals, chemicals, paper and printing, metal products, electronics, and dairy products. Here we evaluate basic metals and chemical industry. We analyse both the quality of the Company Environmental Plans and the degree to which the targets for 1995 have been reached.

Basic metals
Basic metals was the first branch of industry to sign the new style of covenant. It was more or less a test case. The basic metals industry comprises 36 medium-sized to large companies that manufacture metals and metal products. In order to be able to negotiate with the government, a new organization was set up, the Foundation for Basic Metals and the Environment, sponsored by the member companies. The negotiations over the IETP started in 1990 and were concluded in 1992 with a covenant. The covenant was signed by 33 companies. The five biggest ones were together responsible for 80 percent of the emissions. The target plans that were adopted with the covenant formed a direct translation of the national policy aims. Thus, no compromise was made. The main topic for negotiations was the time frame. Taking 1985 as the baseline year, targets were set for the years 1995, 2000, and 2010. The selection of 1985 as the baseline implies that most companies had already made some progress along the path toward the first reduction targets.
Most CEPs were formulated and approved in 1994. The following conclusions may be drawn regarding the quality (see also Biekart, 1994; Boulan, 1994).
- Over half of the CEPs are absolutely no help in clarifying whether or not the state of technology has been attained.
- Hardly any of the CEPs analyze the technical and/or economic bottlenecks that obstruct the achievement of the IETP, nor do they discuss how the various parties intend to deal with these bottlenecks.

- Virtually no CEP makes a connection between the proposed program of environmental measures that should be taken, the costs that would be incurred, and how the program would affect the economic position of the company.
- The CEPs distinguish between definite, conditional, and indefinite measures. These concepts are subject to widely divergent interpretations.
- The measures that were included are almost exclusively those on which prior agreement with the government had been reached (for instance, in the course of the licensing track).
- The CEPs lack a long-range vision. The firms do not give any insight into how they intend to respond to topics such as prevention or clean technology. Furthermore, they do not provide any information on relevant topics such as the development of environmental care systems.

We can now also draw some conclusions regarding the targets realized. For basic metals 1995 targets were set for 28 substances. Ten of these targets are not reached, 8 targets are very well reached and of the other substances the results are not known due to lack of earlier emission data. The overall picture is somewhat more promising when all emission reductions are summed up. Then, it proves that roughly 80 percent of the planned reduction was achieved. Yet a different picture emerges when these results are compared with the estimates made -by both the government and by the private sector- before the target-group policy was started. Assuming that the current policy would be carried out and taking into account the opportunities existing within the prevailing state of technology, the discharge into water bodies was expected to be lower than it proved to be in 1995. With regard to the projected reduction percentages for emissions into the air, in contrast, there was some acceleration. It should be mentioned, though, that the targets related to air pollution were not reached on a number of elements.
The lasting impression is that no systematic search has been made to identify potential measures that would most likely be possible in light of the current state of technology.

Chemical industry
This is a larger branch of industry, with over 100 firms, including some very big multinationals. Together, these 100 or so firms account for 40 percent of the environmental pollution produced by industry. The covenant was signed in 1993. All important firms participate. Like the basic metals branch, described above, the IETP for the chemical industry has not compromised the national aims of environmental policy. The quality of the CEPs is slightly better compared with basic metals. Although there is a lack of real priorities based on analyses stemming from the state of technology. Then too, a comprehensive assessment of

potential emission reductions, taking into account the financial-economic options, has hardly been made explicit. The firms set their sights on the near future, pursuing short-term solutions. They do not explicitly deal with strategies for the long range; to do so would entail more fundamental solutions (that is, source-oriented and process-integrated solutions). In contrast to the situation in basic metals, the chemical industry is pretty well on track in achieving the targets. There were 43 targets set for 1995; 39 of these have been reached for at least 95 percent. For the year 2000, 73 targets have been formulated. It seems that 67 of these will most likely be achieved.

Basic metals and chemical industry compared
The chemical industry is better poised than the basic metal industry to achieve its targets on time. As pointed out above, this does not reflect upon the quality of the CEPs. In both sectors, the quality leaves something to be desired. Furthermore, the CEPs in both sectors contain few new measures. Thus, to explain the difference in progress, we should look first of all at the economic position of these sectors. Basic metals were worse off than chemicals, and the market for basic metals has shown only gradual improvement. The techniques to achieve the reduction in emissions do exist. But the financial problems involved in applying those techniques are greater in basic metals than in chemicals. Secondly, we should take account of the fact that for a long time, the chemical industry has been actively engaged in dealing with environmental issues. To some extent, the industry has been forced to do so by problems with its image. For this branch, the target-group policy represents less of a break in the trend than it does for the basic metal industry. A third reason for the difference in progress between these two sectors is organizational. For quite some time, the chemical industry has had an active branch organization that kept in close touch with the membership. That contact also concerned environmental issues. In contrast, the basic metal industry was less well organized. They still had to set up a suitable structure for internal communication with regard to environmental issues.

7.9 Interpretation and evaluation of the research findings.

How should we interpret the situation sketched above? First, it should be recalled that when the initial covenants were arranged under the target-group policy, only a framework was given. The content of the policy strategy was supposed to be determined in the course of the process. For instance, it proved necessary to make instruction manuals on how to set up a CEP as well as on how to evaluate a CEP. In addition, the provinces still had to set up their internal organization in accordance with the new working methods. With this in mind we can conclude that the companies, but also the government authorities charged with oversight, have limited their input in the first round of CEPs to creating a system out of measures that had already been proposed or that were self-evident. The parties took the path of least resistance. In a sense, the intended route, whereby agreements from the CEPs were to be transformed into permits, was traced backwards. Previous agreements, cemented in a permit, were transformed into the CEP. For companies with a current permit, this did not create many problems. The weakness of the CEPs shows up precisely at the core of the target-group policy, namely its comprehensive quality, the application of the state of technology, and fitting environmental measures to the company's economic situation. For the basic metal branch, it turns out that there was not a sufficiently accurate picture of the options that were possible, given the state of technology. The government bodies involved did not pick up on the opportunities that were offered in this area. This was also the case in the chemical industry. However, there the targets could be attained without much trouble. A crucial point that applies to both branches of industry is the evaluation of the program of environmental measures in relation to the economic position of the company. Governments lack the knowledge to pass judgement on the economic feasibility of the measures at that level. Yet it remains unclear what kind of information would be required to do so. The assessment criteria have not been formulated. Even though a more systematic approach is currently being developed, it is still important that the government is almost entirely dependent on the information that a company itself provides. These critical remarks notwithstanding, the target-group policy has performed positive functions.

- First of all, the consistency of government policy has been enhanced by coordination of the activities geared to a branch of industry. The Ministry of the Environment and the provinces act more in concert and have adapted their internal organization accordingly. In fact, the role of government authorities in charge of granting permits has been written into the national strategy for environmental policy.

- Secondly, within the industrial branch, a similar type of convergence is going on. The branches of industry have taken it upon themselves to achieve the environmental targets. Companies have been motivated to communicate with each other about how to tackle the problems they face. They have been encouraged to develop ways to resolve them jointly.

- A third positive effect is that a dynamic approach to environmental problems has become visible within the companies. Especially the large companies have gained from this policy. The target-group policy has led to broad internal discussions about how to develop a more systematic approach to pollution abatement.

- A fourth effect is the creation of a much more transparent picture of the environmental load of individual companies -previously confidential information- and of the branch of industry as a whole.

- Finally, it is important to note that the policy has had a more general effect. It has led to a deeper understanding of the environmental problems and their possible solutions, as well as to insight in the scope of the task. In this regard, it has become clear that restricting the emissions in question to agreed levels is merely a technical problem. The environmental issue -to the extent that emissions are involved- is primarely a financial and economic problem, which takes time to resolve.

The target-group policy has increased the support for environmental policy. It has generated commitment to the long-term national environmental targets. The number of companies that do not participate -the free riders- is low. They can be forced to comply by way of the licensing track. Thus, there is definitely a learning process under way. Given the structural and processual elements that have been introduced, that learning process can be pursued even more vigorously. And indeed, that seems to be what is happening, now that the aggregate figures on the CEPs are being analyzed by the industrial branches. Their assessment should help determine the next steps, directed at the target plans for the year 2000.

7.10 Conclusions

The environmental covenant is no cure-all. At first, the instrument was mainly used as a marketing device in propagating environmental policy. Since then, it has turned into a full-fledged new concept in governance. In this process government and business gradually learned how to handle this instrument.

Initially, the relations between the government and the private sector were essentially antagonistic. In conjunction with the environmental movement, the government more or less coerced the parties to enter into a covenant. Nevertheless, this phase does contribute to the application of the environmental covenant in a later stage. It is crucial for the government and the private sector to discover that it is indeed possible to make reciprocal arrangements on the effort to achieve the aims of environmental policy.

In the following phase, the character of the negotiations between the public agencies and the private actors is subject to change. At that point, the interaction is for the most part cooperative. The second phase also differs from the warming-up period in other respects. For instance, the commitment of the environmental movement declines as the covenants take shape. In fact, the movement becomes mainly critical of the use of this instrument, which they see as a flimsy tool. And they have good reason to be wary in that phase. There is great diversity among covenants. Whereas some have a good track record, other covenants have definitely missed the mark. The discussion has therefore turned to the legal aspects. At that time the prevailing opinion is that the efficacy of an environmental covenant depends largely on the legal guarantees that can be provided to back up the agreements. However, it is gradually becoming clear that the strength of the instrument may lie in its malleability. In formulating a covenant, the parties might well choose not to make it watertight. It is most important that the parties know which aspects should be included in the agreement and that they are clear about what they do and do not want to put in writing.

When environmental covenants are encapsulated in a broader policy strategy -the target-group policy- during the third phase, the legal aspects are actually pushed into the background. The new environmental covenants stipulate that the agreements are enforceable under civil law. However, the parties only commit themselves to take certain steps in a policy process. With regard to the content of the policy, they are only required to make an effort to attain the targets.

The pivotal question in environmental policy has always been how the collective advantage of a clean environment can be linked to the interests of the individual company. At the beginning of the 1990s, however, this question took on a new urgency. The first projects to clean up pollution have been carried out. That experience has demonstrated that if we are to embark upon a course of sustainable development, there will have to be deep cuts in the emission profile of industry across the board. It is almost self-evident that the changes in the character of the policy aims will lead to a quest for a new strategy of governance. It is imperative to initiate a long-term effort, which hardly seems possible solely on the basis of legislation. The government and the private sector are deeply dependent upon each other in their efforts to achieve the environmental aims. That reciprocal relation is not considered to be a problem, however. The parties are able to put a positive spin on the situation. They build their image of reality upon the assumption of shared interests. They share the ideal that the environment and the economy should go hand in hand. Accordingly, it becomes possible to make the dependencies productive for environmental policy. In the new strategy of governance, companies are thus no longer exclusively seen as objects of regulation. Rather, they are perceived as active participants in a process geared to environmental change. The governance can be directed toward the creation of institutional provisions for a process that would give shape to the joint responsibility of public and private parties for the attainment of the environmental aims. In this way, a model of co-operative management coalesces.

The intent to change is part and parcel of the national aims for environmental policy. In the model of co-operative management, that intent is transformed into an organizational issue. The environmental covenant fits into that model as a fixed point in an ongoing process. At that point, the parties confirm that they have reached agreement on which subsequent steps to take. This model can be applicated if some prerequisites are fulfilled. An environmental covenant can be closed if it is possible to identify specific groups with a common interest. Those groups should also have an effective internal structure and a certain authority. Furthermore, it is important to have a balance of power between public bodies and private parties. These requirements also represent the limitations of governance by way of environmental covenants. An equilibrium in the balance of power is attained through the common standpoint that the environmental policy must be not only technically possible but also economically feasible. On these grounds, the intent to change is essentially restricted. In this regard, Biekart (1996) speaks of an environmental issue that is translated into a management problem of limited proportions. This is expressed in the items that are and are not regulated in covenants as we know them now. Attention is focused mainly on the reduction of emissions of contaminating substances. Out of all existing solutions, industries learn to select the most environmentally friendly alternatives. Other issues impinging upon sustainable development either do not fit into an environmental covenant or would be hard to deal with under that type of arrangement. To name a few, these include the desire to restrict the use of natural resources; the process of thinking through product development with respect to environmental impact in its widest sense; the search for alternative sources of energy; the introduction of product responsibility throughout the entire life cycle of a product; and so forth. Yet to move forward in the quest for sustainable development, more innovation will be necessary. Covenants dealing with these topics still need to be concluded and may become the next 'co-operative' step.

8. AN ECONOMIC APPROACH TO ENVIRONMENTAL AGREEMENTS

Experiences from Germany

HEIDI BERGMANN, KARL LUDWIG BROCKMANN, KLAUS RENNINGS

8.1 Introduction

National and international environmental policies are moving towards 'softer' regulation, especially in the form of (negotiated) environmental agreements. Supporters of this trend emphasize the positive effects on the economy and the environment, calling these agreements 'efficient', 'innovative' and 'environmentally friendly'. The German Federal Government puts its faith in these positive effects, espousing environmental agreements in principle. However, by giving preference to these 'soft' instruments, the government may be tacitly relinquishing its responsibility for the environment. Negotiated agreements made by the business community -like the pledges to develop so-called three-litre cars, to take back old cars, or to reduce carbon dioxide emissions- are therefore controversial from an ecological and economic point of view.

Here, the economic effects are described from a neo-liberal perspective. According to the liberal approach, markets should co-ordinate the allocation of goods as long as they do not disrupt the working of the market. But when external effects, information deficiencies, or inflexibilities arise, the state has to take action. However, the state has to take the option that minimizes market distortions. In other words, market-based instruments like taxes or tradable emission permits are generally preferable. Nevertheless, other types of instruments -like environmental agreements- can be used, either as an alternative or as a supplement, when they offer some advantage (e.g., efficiency, institutional controllability, minimization of unwanted side-effects).

The German Ordo-Liberal School, in the tradition of Eucken, argues in favour of a government that corrects market failure and against an interventionist state. The principles and criteria of the Ordo-Liberal School are outlined below. There, we construct an analytical framework for the valuation of environmental policy measures, giving special attention to environmental agreements. Since Ordo-Liberalism is mainly a German school of economic thought, it should be mentioned that similar ideas have been advanced in other countries. For example, the

P. Glasbergen (ed.), Co-operative Environmental Governance, 157–177.

Oxford Liberals have formulated the adage of "as much competition as possible, as much planning as necessary" (Grossekettler, 1991 p. 106).[1]

8.2 General characteristics of environmental agreements

The term 'environmental agreement' stands for a bundle of instruments. An agreement between industry and government can be tailored to the situation. It may set goals or spell out technical instructions for a specific industry, but it may also stipulate scales of charges or a format for information systems. That is why the agreements do not constitute procedures in their own right. Rather, the mechanism by which the aims are to be achieved depends on the instruments provided in each specific case.

In Germany, environmental agreements are generally said to be 'voluntary'. Actually, the situation is more like 'showing the instruments of torture to the victim as a first step of ordeal'. Comparable to what, in former times, tormentors would do with their tools, the Minister for the Environment presents a draft for an ordinance to extract 'voluntary' concessions (Murswiek, 1988, p. 985). The government uses the 'threat' of enforcing restrictive legal norms as a starting point for negotiating an agreement. In this manner, the authorities exert a guiding influence on the business community's conduct (Hoffmann-Riem, 1990, pp. 400, 426; Brohm, 1992, pp. 1025, 1027). Hence, such cooperative solutions can be interpreted as barter transactions. The business community thereby obliges itself to act in a certain manner, and government in return refrains from coercing them to display the desired conduct.

In the debate on environmental policy, negotiated agreements are sometimes pictured as 'a free-market instrument'. This classification is only justified when the designated instruments are based on the price mechanism. In fact, this is more often the exception than the rule. There are only a few environmental agreements that are founded upon this free-market principle. One such agreement is known as the 'Grüne Punkt' ('Green Dot'). In economic terms, it can be characterized as a levy, whereby the costs related to the disposal of packaging are charged in accordance with the 'polluter pays' principle. As a rule, environmental agreements shy away from such 'tough' economic instruments and do not meddle with the price structure. After all, environmental agreements do not spring from the market system; they are the result of political negotiations between government and trade associations. However, an approach that is essentially

---

[1] However, the Ordo-Liberal School focuses more on the responsibility of the state to establish basic rules for a regulatory framework within long-term economic policy, while the Oxford Liberals give more weight to policy measures in the short run.

based on negotiated solutions should not be characterized as a market-based approach but as a corporatist approach (Holzhey/Tegner, 1996, pp. 426-427). The main difference is that consumers rarely participate in the negotiation process, although the consumer surplus is strongly influenced by the outcome (improvement of environmental quality).

8.3 Negotiation and enforcement of environmental agreements

Environmental agreements between the state and industry differ markedly from the classic model of 'command and control' (cf. Glasbergen, chapter 1). Government refrains from using any instruments of formal power, i.e. from enacting legal norms. Instead, the authorities enter into negotiations with the groups affected on the realization of the environmental goals. In contrast to traditional sovereign actions, government seeks a solution via consensus-building. However, this bargaining process has some pitfalls. Agreements are based on a system of 'reciprocal service'. On the government side, the price of an agreement is often a reduction in the level of environmental protection. In order to reach consensus, government accepts 'a downward adjustment of goals'; the case studies reviewed here suggest that there is always a 'decrease in the stringency of regulations'. The upshot is a tendency for agreements to fall short of the opportunities for implementation provided by laws and ordinances (Müggenborg, 1990, pp. 909, 915). The reasons to relativize the level of environmental protection in connection with cooperative solutions are complex. In itself, 'agreement' suggests a shift away from maximum demands, since a joint solution consists of a package of regulations, a combination that is deemed acceptable to both sides. Inevitably, the readiness of the business community to negotiate only goes as far as their economic self-interest. From their perspective, protecting the environment is not a value in its own right. Instead, environmental values have to coincide with the companies' own (economic) interests. Finally, it should be kept in mind that it is not individual businesses that negotiate with government. As a rule, they are represented by their associations, which -as experience shows- tend to take a more rigid stance.

Concerning the enforcement of an agreement, it should be stressed that a mere promise is no guarantee of implementation. Companies not complying with the pledges (made by their association) run no risk of any fines, penalties or other coercive measures (cf. Hoffmann-Riem, 1990, pp. 400, 438; Murswiek, 1988, pp.985, 988). An environmental agreement entails no performance obligations. Indeed, what characterizes agreements is the lack of any binding power or enforceability. In contrast to a contractual relationship, the party pledging voluntary measures never enters into any legal obligations. Contracts and agreements are similar in that both pertain to an exchange of services. The relationship of 'do ut des', a mutual give-and-take, is typical of contractual relations and agreements alike. However, the binding force of a contract, 'pacta sunt servanda', is completely foreign to environmental agreements. They are more likely to be categorized as a gentlemen's agreement, which does not entail any legal consequences. Even if the government were to be granted the right to monitor compliance with the agreement, this does not constitute a claim on performance. In fact, the point is that the businesses subscribing to the agreement do not want to be bound the way they are by a contract. The companies (or rather their associations) make an agreement without the intention of legal consequences. The agreement consists solely of a declaration of the companies' willingness to act in a certain way, tolerate something, or refrain from doing something. The government side is not granted any claim to performance. Consequently, any legal action founded on the agreement brought against the signatory party would be doomed to failure. Whether or not compliance with the agreement is monitored makes no difference from a legal point of view, since violation of the agreement does not entail any legal consequences (cf. Beyer, 1986, pp. 277 ff.; 282; Henneke, 1991, pp. 267, 271 ff.; Brohm, 1994, pp. 1025, 1034).

8.4 Analytical grid for assessing environmental agreements

The Freiburg Ordo-Liberal School holds that the absence of markets, and any functional defects of existing markets, will prompt corrective requirements which cannot be covered solely by an evolutionary competition of institutions. This view will be elucidated below. That school regards as inadequate the process of natural selection emerging from international competition between institutions; these entities are formed in a quasi-random process, but not specified by the state, which results in the elimination of inefficient institutional bodies. On the contrary, the Freiburg Neo-liberals' concept of a free-enterprise economy presupposes that the state must selectively create institutions to counter existing defects and challenges (Grossekettler, 1991, pp. 104-106). The efficiency of such institutions must be measured in terms of how far they uphold the principles of free enterprise.

Seen from the perspective of Ordo-Liberalism, environmental agreements are regarded with skepticism. Government no longer -as postulated in the Ordo-Liberal model- stakes out the regulatory framework within which entrepreneurs can dedicate themselves to profit-making. Instead, government delegates this responsibility to the businesses themselves. However, in the opinion of Ordo-Liberals, a laissez-faire policy, with a market economy left to its own devices, tends to destroy itself. Failure to protect and promote competition may lead to closed-off markets, cartels, price-fixing, and ultimately to a distorted price structure. A lack of competition and the wrong price signals, in turn, may provoke a spate of government interventions. For instance, the state may take measures to monitor prices or adopt regulatory requirements concerning the application of given environmental technologies (due to a lack of price incentives). That is why those economists who consider government to be a powerful custodian of the system in the tradition of Eucken's Ordo-Liberal School view the trend in environmental policy towards environmental agreements with great concern (Maier-Rigaud, 1995).

However, the following assessment will test some of these hypotheses by applying the Ordo-Liberal principles to concrete case studies of environmental agreements. Based on a broader set of criteria (Grosskettler, 1991, p. 114; Rennings et al., 1996) the following questions were formulated.

- *Goal conformity*: Is it possible to achieve the specific underlying environmental policy goal with the help of the given agreement?
- *System conformity*: Is the measure in keeping with the system of a social market economy? Is the system strengthened, does it remain unchanged, or is it weakened? Are detectable unwanted side-effects minimized?
- *Economic efficiency*: Are the costs higher or lower than when using other instruments? What incentives are provided to achieve technological progress?
- *Institutional controllability*: How immune are environmental agreements to political influences that can dilute their desired effect? How do the associations deal with the problem of free riders?

8.5 Choice of case studies

The assessment questions to evaluate environmental policy measures are applied to topical examples taken from the fields of climate protection and circular economy. The task of promoting clean technology is cross-sectional in nature because it is relevant for both fields of environmental policy.

Climate protection
The chlorofluorocarbon (CFC) phase-out and the reduction in carbon dioxide emissions will serve as case studies in the area of climate protection. The discussion about the depletion of the ozone layer led to a first agreement in 1977 (followed by a second one in 1987) amongst the Aero-sole-industry to drastically reduce the use of CFC. Chemical industry followed with an agreement in 1990 and in 1992 the hard form industry obliged themselves to do without CFC. The CFC case study derives its relevance from the fact that the progress made in

phasing out CFCs is often cited as evidence of the economic and ecological advantage of using environmental agreements. Agreements, the argument goes, help attain the ecological goal more quickly, they trigger an innovation drive in industry, and safeguard industry's export capability. The carbon dioxide case study is an obvious choice, and not only because of its prominent role in the global warming problem. Accounting for 60 per cent of the damage, anthropogenic carbon dioxide emissions are the largest cause of the greenhouse effect. With regard to the carbon dioxide problem, the German business community presented an updated environmental agreement in March 1996; 19 branches of industry encompassing sugar producers as well as steel manufacturers agreed to reduce their specific energy consumption and their carbon dioxide emissions up to the year 2005 by 20% compared to 1990. That negotiated agreement has come to be a model for the European Union's climate policy.

Circular economy
Particularly in the field of German waste disposal policy, environmental agreements predominate over other measures. The significance of negotiated agreements in the area of waste disposal policy is explicitly stressed by the federal government. In order to place greater emphasis on preventing waste, Germany created the *Kreislaufwirtschafts- und Abfallgesetz* (Act on a Circular Economy and Waste), which came into force in October 1996. Producers of, say, electronic goods and batteries can pre-empt industry-specific ordinances through negotiated agreements and can voluntarily comply with the legal obligations. The automobile industry signed an environmental agreement on the disposal of old cars, which is being used as a case study. With this agreement the German automobile industry grants to build up an infrastructure for the recycling of old cars and to take back their own products for free provided that the cars are in good shape and not older than 12 years. The obligation is only valid for cars registered after the agreement came into force. In addition, the *Duales System Deutschland* (DSD, Dual System Germany), founded in 1990, will be evaluated here. The DSD established a takeback system for packaging materials organized by trade and industry.

Clean technology
A further case study to be examined here is the so-called three-litre car, which serves as an example of clean technologies. In March 1995 automobile industry obliged themselves to reduce the average consumption of fuel by 25% until 2005 measured by the standard of the year 1990. There is some overlap between that case study and the other ones. Clean technology is used in climate protection as well as in strategies of material flow management aiming at establishing a circular economy. Nonetheless, it is instructive to deal with this area separately. It seems likely that in the future clean technologies will be of great importance as we move

along the road to a sustainable economy. From that perspective, the question arises how such a development can be promoted without repeating the mistakes made in the past (regulatory enactment of best available technologies) and what part environmental protection measures can play in this development.

8.6 Goal conformity

CFC negotiated agreement: in certain applications in conformity with the goals. So far environmental agreements to reduce ozone-depleting substances have definitely been successful in ecological terms. The goals were surpassed, although some critics felt the phase-out could have been achieved even faster (Kohlhaas/Praetorius, 1994 p. 89). As a rule, the negotiated agreements are monitored by someone who is neutral. Negotiated agreements governing the reduction of CFCs enjoyed special advantages (compared with the case study on the reduction of carbon dioxide). One advantage is that substitutes had already been discovered which, when applied, did not lead to soaring costs. Another is that additional pressure was exerted by the demand side (causing a slump in sales of sprays containing CFCs). Thus, in terms of an entrepreneur's cost-benefit calculations, opting for a reduction in CFCs involved little risk. On the other hand, continuing to manufacture products containing CFCs would have been much riskier from a managerial point of view. This conclusion is borne out by the fact that although CFCs were speedily replaced in products, CFC substitution in production processes was slow. Manufacturers were hesitant to move ahead because the switch was costly and there was less pressure by the demand side.

No impetus for absolute reductions in CO_2 emissions
The business community reached agreements on a reduction in carbon dioxide in 1991 and 1995. These agreements came in for a lot of criticism, because they did not contain any noticeable initiatives that clearly went beyond 'business as usual'. Criticism was levelled in particular at the fact that it was hard to check whether the goal was being achieved. The agreements did not meet the following minimum requirements for information (listed by the German Federal Environmental Protection Agency) that should be included in a negotiated agreement on carbon dioxide reduction (Ökologische Briefe, (2), 1966):
- reference and target year as well as a reduction path in the form of a timetable with detailed information on partial goals;
- exact fixing of emission or energy conservation goals;
- absolute energy consumption listed according to fuel;
- development of primary consumption;
- development of specific consumption per technical unit;

- reductions achieved;
- in-depth comment on and analysis of the figures provided (e.g. information on whether reductions are attributable to additional climate protection activities, an economic slowdown or to modernization investments that would have been made anyhow); as well as a
- detailed list of the 'special efforts' promised.

The updated version of the negotiated agreement of March 1996 does meet these minimum requirements. The plan on carbon dioxide monitoring presented by the BDI (Federation of German Industries) provides for associations to record their reductions in a total of eight tables (BDI 1996). Total fossil fuel input, net power supplied externally, energy input, as well as specific carbon dioxide emissions calculated from this data and specific energy input are to be stated for the base year, the previous year and the year under review. The requested comment on and analysis of the figures and the list of special efforts are also included. The reports are checked by a neutral expert.

What is particularly striking is that no fewer than 12 out of 19 associations pledge to reduce absolute carbon dioxide emissions. According to the Federal Government (Bundesregierung, 1996a, p. 3) the pledges correspond to a 20 per cent reduction in emissions in these sectors. However, on closer examination, it is precisely the achievement of absolute reduction goals that turns out to be the agreement's real Achilles heel. This is illustrated in a statement made by the *Vereinigung Deutscher Elektrizitätswerke* (VDEW, Association of German Electric Power Stations) (VDEW, 1996).

According to the VDEW, the absolute reduction potential in the electricity industry until the year 2015 amounts to 25 per cent compared with the base year of 1987. However, compared with 1990, which the Federal Government is now taking as a basis, estimates put the potential at a mere 12 per cent. The discrepancy is even greater if the target year of the Federal Government is taken as a basis. The potential for the period up until the year 2005 is, compared with the reference year of 1990, only eight to ten per cent. Compliance with these pledges is even conditional on ambitious prerequisites such as:
- a consensus in society on regarding exploitation of nuclear energy on the basis of existing law;
- an increase in the service life and capacity of existing nuclear power stations;
- the Mülheim-Kärlich nuclear power station going into operation;
- the undisturbed operation of existing nuclear power stations; and
- unrestricted choice of fuels for the power stations on the part of the companies.

On the whole, with regard to goal conformity, it is clear that the instrument of environmental agreements has some shortcomings, and these have been identified and in part abolished. Still, when a comparison with the Federal Government's climate protection goals is made, one has to question the goal conformity of this instrument, especially if all other measures are superseded by the agreement that was made. To date, the development of absolute carbon dioxide emissions has by no means taken a course that makes this goal seem feasible (Kohlhaas/Praetorius, 1994, p. 278). Nor does the updated declaration issued by the German business sector suggest a new trend. In that light, there seems to be a need for action, particularly in the German states that made up the FRG prior to unification. Consider the fact that the VDEW estimates the absolute carbon dioxide reduction potential of German electric power stations, which are responsible for about a third of all German carbon dioxide emissions, at a mere eight to ten per cent, compared with the target year of 2005 (VDEW, 1996, p. 5). That estimate gives rise to the question of who is to make the above-average reductions needed to offset the expected increases in carbon dioxide emissions in areas like transport.

Takeback of cars: business as usual
With regard to environmental agreements on the disposal of old cars, the assessment of goal conformity is focused on whether the selected instruments are the right ones. Are they suited to achieve the main goal, to reduce solid waste production in general and that of shredder waste in particular? First of all, the substantive contribution in the automobile industry's agreement should be compared with the way in which old cars have been disposed of so far. If some substance can be established, the next question is whether the planned measures would really help resolve the problem.

Against the background of the conditions in the existing auto-wrecking industry, the measures are not in keeping with the goals. They do not represent any departure from the status quo. A nationwide takeback network is already in place, made up of independent businesses. Besides, even today, the last owner is able to have his or her old car disposed of at the going market rate. Thus, the last owner still manages to get a good price for the old car.

The pledge to take back old cars at no charge as they become redundant in the future car fleet -provided they are no older than 12 years at the time- does not constitute a substantive measure either. First, most of the cars that have to be disposed of do not fall into this category. According to the ADAC (the main association of motorists in Germany), the mean age of a car whose registration has been cancelled is 13.2 years. Using data provided by the manufacturers as a basis, no fewer than eight out of 23 producers say that the average life of the vehicles exceeds 15 years. Secondly, the pledge of taking cars back free of charge is obsolete. According to the *Arbeitsgemeinschaft Deutscher Autorecyclingbetriebe GmbH* (ADA, Association of German Car Recycling

Businesses Ltd.), all old cars fitting the definition still fetch a good price (Autorecycling, issue 1/96).

The bottom line of the study is that the essential components of the takeback agreement are 'business as usual' measures. That is, they will not lead to a substantial change in the way things have been regulated so far. It is alarming that the negotiated agreement encourages manufacturers and importers to assume control of the auto-wrecking industry. Here, a paradox may arise. Due to the environmental agreement, the polluters -i.e., the producers with whom product responsibility lies- will not be charged costs in line with the polluter-pays principle. In fact the polluters are even provided with a framework for opening up a novel and lucrative business segment.

On the whole, the agreement to recycle old cars -as signed by the German automotive industry and its key suppliers, the car component distribution sector, and the car salvage sector- can be characterized as a 'business as usual' measure. Thus, the environmental policy goal has only been met to a very small degree.

Clean technology: policy mix including economic instruments makes sense
Efficient measures to promote clean technology should take into account a company's investment cycles and adjustment deadlines. In principle, environmental agreements can support this process, but by themselves they are often unsatisfactory with regard to desired targets.

Agreements presupposing a commitment to refrain from introducing auxiliary economic instruments, in turn, have to be considered inappropriate to the goal. A case in point seems to be the meeting of three German heads of government with representatives of the automotive industry in March 1995. At that meeting, moderation was pledged concerning additional government measures such as an increase of tax on fossil fuel. However, it is precisely the pledge to refrain from adopting these accompanying measures that undermines the prospects of marketing an energy-efficient car successfully.[2]

With respect to the standards of vehicle fleets, it is unknown whether the goals will be achieved, even at a company level. Compliance with the standards depends on the demand for individual models, which is something a company finds hard to anticipate. So far, all fuel-efficient models have been vehicles with diesel engines. But even modern diesel engines fitted with soot filters emit three times as many carcinogenic substances as petrol engines. Consequently, the

---

[2] The European Commission, which wants to see negotiated agreements on a reduction in the fuel consumption of the motor vehicle fleet, also seems to have realized this. The Commission's estimate is that such an agreement would entail the risk that the newly developed vehicle models fail in the market. That is why the Commission believes such a measure has to be combined with fiscal incentives for the consumers.

German Council of Environmental Advisers warns of a one-sided policy on the reduction of greenhouse gases. Instead, the Council calls for a greater reduction in carcinogenic substances.

8.7 System conformity

Non-binding agreements: no sanctions against unfair players
Environmental agreements can constitute solutions in keeping with the system, if they establish binding standards for the parties involved and if free-rider behaviour can be prevented. However, associations normally do not take punitive measures against their members, and hence no binding rules of the game can be set. Thus, the approach to problem-solving shifts to the moves of the game, that is to say to the wrong level in the system. As Kreuzberg (1993, p. 308) writes, environmental agreements hinge on "a disproportionately high participation of honest companies in environmental protection activities and therefore lead to a redistribution of burdens to the detriment of 'honest' players".

CFC phase out: 'soft' instruments appropriate for preventive strategies
Prohibitions are deliberate measures to avert concrete and acute damage to the environment. These are instruments that are in keeping with the system. In such situations, the liberal principle of maximizing freedom cannot be used as a yardstick for policy design on constitutional grounds (Brockmann et al., 1995, p. 70). That means that instruments with a low intervention intensity and a high degree of freedom, such as negotiated agreements, can only take precedence as long as safeguards are included to avert acute environmental damage. Otherwise, it will be necessary to apply 'tough' measures. This is also evident in the CFC example. The realistic alternative to a voluntary reduction in the production of CFCs and ozone-depleting substitutes would have been a quick ban on these substances. In order to bring about a complete phase-out of CFC application in products, finally the complementary CFC-halon-prohibition ordinance was issued in Germany. The aerosol industry did not agree to a voluntary phase-out desired by the Ministry for the Environment (Kohlhaas/Praetorius, 1994, p. 89).[3]

---

[3] One way to protect the ozone layer that is more in keeping with the system than anything else would be to apply 'tough' measures that offer economic incentives, as was done in the US. There politicians succesfully opted for the application of a mix of instruments comprising levies and permits (Cook, 1996, p. 4).

Climate policy: long-term danger of intervention spiral
One characteristic of environmental protection measures in the field of climate protection has frequently been cited as particularly appropriate to the system. Compared to energy and carbon taxes, these measures have fewer side-effects on goals relating to stability and distribution policies. The 'double dividend' of an ecological tax reform is often overrated, or so the argument goes. In other words, too much is expected of the simultaneous achievement of ecological (reduction in greenhouse gases) and economic goals (e.g. job creation) due to the revenue-neutral compensation of ecological taxes (Koschel/Weinreich, 1995). However, if emitters are charged the adjustment costs incurred by a climatologically sound restructuring of capital assets, this has to be regarded as being in keeping with the market, irrespective of the compensation question. A 'policy of little steps' which the German Council of Environmental Advisers advocates, would primarily send the necessary price signals to consumers for a more economical use of energy. This would be the crucial advantage over environmental agreements. Secondly, a policy of little steps would minimize the side-effects on goals in the fields of stability and distribution policies. Environmental agreements are basically at odds with the market and the system, because such price signals are seldom included in the instruments they are provided with. A climate policy based on agreements is thus always associated with the danger that even a high 'no regrets' potential that could be siphoned off is overcompensated by increasing energy consumption and not only in the area of transport.

In the long run, failure to attain the goals leads to new government intervention in the form of restrictions, technical instruction, and rules of conduct. As long as there are no permanent and effective incentives to buy and use energy-efficient technologies or to practise energy-saving behaviour, failure to reach the goals has to be countered with the help of such new regulations. Regulations on driving behaviour, the purchase of energy-efficient motor vehicles, or on using state-of-the-art technologies for heat insulation purposes are examples of such measures. Unless plans for voluntary climate protection measures are ultimately accompanied by price instruments which are in keeping with the market, there is a risk of setting a spiral of intervention in motion that should have been stopped by this instrument in the first place.

Circular economy: effects of negotiated agreement on competition
Co-operation between individual trade associations and the government can lead to informal marketing agreements, to markets being shielded, and to barriers preventing third parties from entering the market. The case studies taken from the problem area of the circular economy show that these fears cannot be dismissed. From the point of view of competition policy, buzz words in environmental policy

like 'product responsibility from cradle to grave' have a hidden power. They encourage trends towards concentration and vertical integration.

Negotiated agreements on the disposal of old cars have an impact on the structure of the disposal industry. It is the effects on the number of intake points and on the number of businesses doing the recycling that are decisive. In general, as standards for the sector become more stringent, trends towards concentration in the industry may be expected.

In principle, all specialized businesses meeting the criteria for recognition or certification enjoy unrestricted access to the system. However, there are dangers involved in the formulation of the criteria licensed businesses have to meet. Barriers to access could be erected, favouring financially strong companies, because large investments prove necessary. The Association of German Car Recycling Businesses (ADA) decries the threat to the small and mid-size scale of the industry (Autorecycling, issue 1/96). Besides, if vehicle manufacturers largely control the salvage industry, the option of recycling might become less attractive to auto-wreckers, as manufacturers will probably prefer to sell new parts.

Competition in the recycling industry is indeed under threat. The small and medium-sized businesses stand to lose the most, owing to cooperation between large manufacturers and suppliers. This can happen at the expense of the hitherto dominant small and medium-sized businesses -a pattern already familiar from the development of the Dual System Germany.

Irrespective of licensing criteria for businesses, a network could, once it has been set up, be subject to the abuse of power. Like the relationship between large motor vehicle manufacturers and suppliers, ominous structures can develop between manufacturers and recyclers (Benzler/Löbbe, 1995, p. 157). The takeback network can also lead to a monopoly on the spare parts market and of shredder activities. For example, the Association of the Automotive Industry (VDA) had originally planned to couple the free takeback with brand loyalty for spare parts. But this point was dropped in the course of the negotiations (Homburger, 1996).

Enhancing cleaner production: fair rules instead of voluntary compliance
End-of-pipe technologies will not automatically be superseded by cleaner production (SRU, 1994, Tz 262). According to the German Council of Environmental Advisers, the prospects for clean technology depend on progress in environmental policy with respect to correcting the structure of relative pricing. That structure encourages overexploitation of the natural life-sustaining systems.

On the other hand, environmental agreements, by virtue of their approach, are often not up to the job. Instead of accompanying a correction of relative prices, they are mainly used as a substitute thereof and signed without giving due consideration to the market.[4] Even if, for example, fuel consumption in motor vehicles can be cut, additional price incentives must be provided. Unless the consumers are induced to use energy economically and to buy these vehicles, these innovations will remain part of the decor at motor shows, without large numbers of them being sold to the public.

As long as the wrong rules apply, it seems unlikely that a player is voluntarily going to behave in the desired ecological manner. However, whenever a desired form of behaviour fails to materialize voluntarily, government can respond in two ways. Either it puts up with failures in environmental policy -thereby showing that it leaves the design of the political framework up to influential pressure groups- or, sooner or later, it helps things along with coercive measures. In this way, even though voluntary measures in themselves have a low intervention intensity, in the long term they may lead to a spiral of intervention with an unacceptably high intervention intensity.

Green Dot: Market-oriented approach with some market imperfections
As for cooperative solutions, it is possible to develop environmental policy instruments that take free-market principles into account. The instrument of imposing dump charges to help defray the social costs related to the dumping of waste proved to be problematic. In contrast, the ordinance governing packaging, for instance, and the Dual System it spawned represent two more promising variants. These instruments of control make sense in economic terms and are in keeping with the market. From a legal perspective, a takeback obligation is a directive. Such enforceable directives are less questionable from an Ordo-Liberal point of view, since they provide scope for individual deviations (Spies, 1994, p. 309).

In the field of packaging, too, some reservations arise from the point of view of competition policy. German disposal markets are highly cartelized and oligopolized. The Dual System is a demand cartel legitimated by the ordinance on packaging. There is always just one disposer per territorial division who is

---

[4] For companies, the switch to clean technology is linked with increasing cost and financial risks in the short term. From the point of view of an entrepreneur's investment calculations, the application of integrated technologies has several disadvantages. Higher access and information costs, adjustement and changover costs, funding bottlenecks, long decision-making horizons and greater economic risk are the main drawbacks. Negotiated agreements may be able to provide an additional impetus for adjustment and changeover. But an adjustment to ecological requirements in a way that minimizes the costs for the overall economy is not to be expected with environmental agreements. Adjustment along these lines is unlikely in view of the restriction to certain technologies and the expected free-rider behaviour on the part of the companies.

given the tasks of collecting and sorting.[5] However, the Dual System also shows that non-uniform action in different countries can lead to distorted competition in foreign trade. Negative impacts of the Dual System on the European disposal industry become apparent in the case of waste paper, for example. German recyclers get their waste paper free of charge, whereas other European recyclers have to bear the collecting and sorting costs themselves. This cost advantage represents a de facto import impediment, but it also leads to increasing German exports of waste paper. For example, in France this has the ecologically ludicrous consequence that growing amounts of waste paper have to be incinerated or put on dumps for lack of demand (Wupperman, 1993, p. 453).

8.8 Economic efficiency

CO$_2$ agreement is not cost-effective
Solutions to the problem of reducing emissions are economically cost-effective if each emitter fixes his contribution to the reduction in such a manner that the overall economic avoidance costs are minimized. Individual emitters neither know the avoidance costs nor are they interested in including them in their decision-making. Thus, these signals have to be sent through environmental policy. Levies and permits in particular are classic textbook instruments meeting the economic efficiency criteria, because they send out these signals in the form of a correction of relative prices. On the other hand, a reduction in emissions with minimal costs for the overall economy is unlikely to be achieved via an environmental agreement on carbon dioxide or eco-efficient cars, for instance. The problem lies in the restriction to certain sectors and the free-rider behaviour to be expected on the part of the members of the associations.

Moreover, environmental agreements are not convincing when it comes to their dynamic efficiency -i.e., their effect on technological progress. In this respect, it is possible to compare environmental agreements to regulations. Once the goal has been achieved, there are no further incentives to reduce emissions. Compared with regulations, there does not even have to be an incentive to maintain the standard once it has been attained (Kreuzberg, 1993, p. 309).

---

[5] The Federal Ministry for the Environment justifies the formation of cartels by saying that new environmental technologies will be developed. In a diametrically opposed view, the Federal Cartel Authority in Berlin regards functioning competition as the essential driving force for the development of new environmental technologies (WELT of 15 April 1996). The amendment to the ordinance on packaging, which is to take effect soon, is supposed to promote competition by stipulating that disposal activities (collecting, sorting, recycling) linked to Dual Systems have to be put out to tender (Umwelt, 4/1996).

Cost-effective CFC phase-out

When assessing the economic efficiency of German agreements on reducing ozone-damaging substances, consideration has to be given to three unusual features in particular that accelerated the substitution on managerial grounds alone:

- the ban on CFCs as propellants for most aerosol products in the US as early as 1978;
- the declining demand for products containing CFCs (e.g., aerosols); and
- the availability of low-cost substitutes.

The economic cost-effectiveness of negotiated agreements on CFCs may be regarded as positive. Each agreement covered the main polluters, which were enabled to implement the reduction at the lowest possible cost.

Concerning dynamic efficiency, it should be recalled that once the goal has been achieved, there are no further incentives for progress in terms of environmental technology. However, with respect to the development of environmentally friendly CFC substitutes, the bottom line is that owing to the aforementioned unusual features, there had already been sufficient incentives for technological progress.[6]

Fundamental distortions of waste markets

Although the distribution of scarce environmental resources takes place in an allocation-efficient fashion via a change in relative prices, waste disposal in Germany is confronted with fundamental market distortions. The capacity problem regarding dumps would theoretically best be solved by charging dump users in line with the polluter-pays principle. That means that the users would pay all internal and social costs that arise. This could be arranged by increasing dump charges. The costs related to dump scarcity, dump operation, and dump emissions feature prominently among the costs not fully covered at present. So far, costs related to the environment have been externalized as much as possible. However, on account of an inflexible scale of charges, dump operators are not able to charge users for all their operational costs (Brenck et al., 1996, p. 1). Inefficient forms of organization, in turn, lead to price increases; on the whole, serious

---

[6] What has turned out to be a particularly efficient instrument in the US to protect the ozone layer is the simultaneous application of permits and a tax for the protection of the ozone layer. The American environmental agency EPA estimates that the administrative costs amounted to only 10 per cent of the administrative work a regulation would have entailed. Furthermore, it was possible to quickly adjust the licenses issued to reflect the modifications of the Montreal Protocol. Whereas in 1988 the cost of halving CFC consumption was put at $ 3.50 per kilogramme, only two years later it was possible to lower these estimates to $ 2.20 per kilogramme (Cook, 1996, p. 4ff).

distortions in the cost structures ensue.[7] In view of the promised increase in recycling quotas, considerable costs are incurred in setting up the disposal system for old cars. As the establishment and maintenance of the disposal system represent sunk costs, there is a growing incentive for businesses to partially recycle the old cars that were taken back, or to recycle the raw materials they contain. Yet, whether or not there will be a drop in the volume of waste to be dumped or in net resource input depends on the relative price for newly purchased raw materials or for newly produced car parts.

The fact that cars are taken back free of charge does justice to the polluter-pays principle, since the producer bearing product responsibility is charged for the disposal costs. Thus, the solution is dynamically efficient. The producer is provided with incentives to look for technological innovations in competition with other producers (Sacksofsky, 1996, p. 103). However, the question arises whether the disposal costs are high enough in relation to other production costs to initiate recycling-oriented research and development.

Green Dot economically efficient

In response to the ordinance on packaging, an economically efficient system was created in the packaging sector, on a voluntary basis, by introducing the Green Dot system. A graduated system of royalties paid by packing material producers and distributors to the Dual System Germany Ltd. shows a tendency of charging costs in accordance with the polluter-pays principle. Thus, the system compensates for the marginal avoidance costs. However, some reservations about the system result from the changed conditions in the disposal and recycling of packaging following the establishment of the Dual System. Strong concentration effects in the disposal and recycling markets make monopoly and oligopoly profits possible. In some cases, transaction costs have risen considerably compared with the status quo ante. The financial crisis experienced by the DSD in 1993 halfway through the year illustrates the level of transaction costs associated with this system. On the expenditure side, collecting and sorting costs had soared on account of higher-than-expected volumes. As for the disposal of the remnants, the problem that emerged was that disposal businesses owned by private persons or local authorities are able to charge monopoly prices. With respect to the revenue side, the Green Dot was frequently used without making the corresponding

---

[7] Shredder residues were classified as hazardous waste in 1990. This measure, aimed at containing shredder residues from the disposal of old cars, had a great impact on prices. Dumping costs rose from around DM 70 per ton to between DM 500 and DM 1,000 per ton (1992), (Weiland, 1995, p. 58). However, it is still too early to comment on the measure's direct effect. According to the Association of German Car Recycling Business (ADA), the share of waste from the disposal of old cars -in partucular, plastics- is still increasing.

royalty payments to the DSD. By September 1993, a deficit amounting to DM 710 million had accumulated (Wupperman, 1993, p. 450).

8.9 Institutional controllability

Enforceability depends on induced costs and external pressure
In principle, the enforceability of environmental agreements in associations depends on three factors:
- the costs of the agreement made;
- the costs of the government ordinance in the event of non-performance of the agreement; and
- the effectiveness of punishment meted out by the associations themselves (e.g., expulsion from the association).
It is unlikely that 'no regrets' measures will encounter much opposition in an association. Yet pledges going beyond them might lead to an internal allocation struggle, which can make withdrawal from the commitment seem worthwhile. In such cases, certain passages in the declaration providing scope for interpretation, or preconditions for the agreement that were never met, can provide a welcome opportunity to terminate co-operation.

Problems regarding enforceability are not expected to emerge with regard to the automotive industry's fairly lenient environmental agreement on reducing fuel consumption. According to the latest annual report of the German Council of Environmental Advisers (SRU, 1996, Tz. 166, translated by the authors): "it is widely believed that the goal of cutting fuel consumption by 25 per cent over 15 years, based on a relatively high average consumption, is in line with the expected technological development already. The pledge to launch car models by the year 2000 with diesel engines consuming three to four litres per 100 km cannot claim to be a special feat of engineering, either".

On the other hand, problems relating to the enforcement of environmental agreements on carbon dioxide reduction in associations have come to light. For instance, enforcement has been discussed in talks conducted by the Federal Environmental Protection Agency (UBA) with representatives from the associations in connection with the climate protection initiative launched by the German business community. In the UBA's annual report (1993, p. 167, translated by the authors), we read that: "The talks revealed fundamental difficulties whenever negotiated agreements made by associations are to contain binding requirements that are usually laid down on a statutory basis. Many individual businesses refuse to recognize declarations made by their associations as binding. It is virtually impossible to impose sanctions following non-compliance of the agreements". Thus, the instrument of negotiated agreements has built-in limits, which result from its voluntary nature.

Redistribution at the expense of third parties

The instrument of environmental agreements per se does not send out any price signals that lead to automatic adjustments on the part of the players. Therefore, decisions have to be made on a case-by-case basis on how the overall reduction targets are to be allocated to individual groups. This, some people say, puts the greatest burden of adjustment upon groups that are inferior in terms of the way they are organized, like households. It should be realized that this argument can also be used against economic instruments. In real life, the political representation of interests also plays an important role with respect to economic instruments. For instance, this is clear when exceptional areas are stipulated that will be exempted from energy taxes. Even Denmark, a country usually regarded as exemplary when it comes to introducing ecological taxes, has special arrangements for particularly energy-intensive businesses. And there, too, the burden of levies is mainly borne by private households (Mez, 1995, pp. 109, 126). As the example shows, the uniform price signals that economic instruments are intended to bring about are frequently watered down in the political process.

Relinquishing political scope

The updated declaration presented by the German business community on the prevention of damage to the climate once again clearly reveals the limits of environmental agreements. If the environmental goal obviously runs counter to the individual economic interests of the associations, people -like some representatives of the VDEW recently- suddenly talk of the 'fetish year 2005' in connection with the Federal Government's timetable. Although the VDEW's agreement is already a scaled-down version (calling for a reduction of only eight to ten per cent by the year 2005), undesired political measures (e.g., in nuclear energy policy) may prompt termination of the agreement. In any case, a government has to expect to be asked whether it is possible for a pledge with little substance to be worth so much that the government in return, for decades to come, will put up with being deprived of a great number of potential courses of action relating to climate and energy policies.

8.10 General conclusions

The case studies analysed here have shed light on the weakness of environmental agreements with regard to goal conformity, system conformity, cost-effectiveness, and institutional controllability. In some cases, it may be reasonable to make use of environmental agreements, but the instrument should be applied very cautiously. Against this background, we can draw some general conclusions from the German experience and offer some points for discussion within the European Union.

From an Ordo-Liberal point of view, environmental agreements can be justified especially if they are used within a mix of policy instruments. An agreement that gives economic incentives like the 'Green Dot' or an agreement that is used to 'accompany' or 'support' economic instruments (for example, the combination of a carbon tax and an agreement) obviously deserves the attribute of 'being market-based'. In cases where economic instruments cannot be applied (as in the complete ban of a substance, for instance), the 'soft' instrument of an agreement can certainly accompany 'tough' regulatory measures or -provided it is not a matter of warding off acute danger- even replace them (for example, a CFC phase-out). In any other situation, however, instruments other than environmental agreements are preferable from an Ordo-Liberal point of view.

Environmental agreements can be accompanied by economic instruments to ensure there is motivation to comply. A lack of incentives is one of the main pitfalls of most agreements. In the absence of incentives, the agreements reached to date have either been undemanding with regard to the content or, in the case of more ambitious pledges, have been associated with a great input of time and effort for enforcement and monitoring. It is also due to a lack of incentives that as a rule, only specific and no absolute reductions are achieved. Even demanding specific reduction goals, without any signals in favour of using environmental resources in ways that are less harmful, there is a real risk that increasing consumption of environmental resources will overcompensate for these goals.

Two factors may explain the increasing significance of environmental agreements in current environmental policies. One is the fact that they have been labelled 'market based instruments'. The other is the fact that their use is associated with strengthening the market system. However, a more in-depth analysis seems to undermine these hopes. In principle, it is possible to design agreements in a manner that is in keeping with a market economy. But in most cases, the parties avoid doing so. As the examples presented here show, a free-market orientation does not come about voluntarily or spontaneously. Rather, it requires standards and a framework set by government. However, once such a framework is set, agreements may be used as a tool for implementation within an environmental policy mix.

For this reason, the government is strongly advised against making a commitment to the effect that in return for 'voluntary' declarations on environmental protection, the government will not make use of any other instruments. This imposes disproportionately severe restrictions on the politicians' latitude in future action and thus on their capacity for problem-solving. Such a policy tends to neglect other solutions that may be more appropriate to the problem, such as the application of a mixture of environmental policy instruments.

It is counterproductive to give preference to 'voluntary solutions' in general or to decide in favour of such agreements at an early stage. These signals weaken the substance of negotiated solutions -the governmental 'potential for threats'- and open the door to delays, testing the perseverance of the parties. The examples that were examined confirm that without considerable governmental pressure, environmental agreements do not yield any pledges that go beyond 'business as usual' or 'no regrets' measures. In order to make it absolutely clear that the government really 'is in control', the scope for design in environmental policy has to be kept unrestricted and flexible. On the other hand, the decision to give priority to environmental agreements may have a disadvantage. It puts government in a bind when it comes to quickly reverting to 'tough' environmental policy instruments following unsatisfactory negotiation results or delayed implementation.

Although our case studies are derived from German experiences, some lessons can be learnt for the use of environmental agreements at the European level. It seems that the European Commission is aware of several potential shortcomings of the agreements reviewed here. Thus, in November 1996, the Commission issued a report on the use of environmental agreements as an instrument of EU environment policy. The (non-binding) paper deals with the instrument in general, presents guidelines on 'environmental agreements', and provides a survey of how agreements are applied in the Member States. It should be noted that, according to the Commission's report:

- a cost-effective use of agreements will be made as a part of the policy mix, together with, for example, regulatory or economic instruments; and
- under certain circumstances, agreements may be an efficient tool for implementing environmental policy, but they are not appropriate for environmental target-setting; general targets should be set through legislation.

These basic conclusions are in line with the findings of our study. The politicians may now have to demonstrate their skill in the 'art of the possible' in identifying reasonable applications of agreements without falling back upon traditional command-and-control measures. As experiences in Belgium (Seyad et al., 1996) and the United Kingdom (Eden, 1996) show, the future does not seem to lie in a strategy of strong regulation of environmental agreements (which makes the agreements obsolete) or in a strategy of far-reaching deregulation (nor in self-regulation) of environmental target-setting. If not driven to one of these extremes, there may be several ways to design more flexible applications of environmental policy.

9. TRADE LAW ASPECTS IN RELATION TO USE OF ENVIRONMENTAL CONTRACTS

ELLEN MARGRETHE BASSE

9.1 A new flexible tool

There is a clear need for further measures to protect and repair the environment as well as to safeguard the welfare, health and life of society. Yet this does not necessarily mean that the focus must be solely on stronger environmental legislation along traditional avenues. A recent shift in the focus of environmental policy, calling for new approaches to problems, moves the emphasis from control over relatively few sources of the most serious industrial pollution to the exertion of influence over a very large number of smaller, highly decentralised and diverse, sources. Each of these may be of relatively little significance, but cumulatively they are of very great importance. Connected to this is a change in governance strategy.

A flexible, co-operative enforcement style -based on a bottom-up approach, self-control, self-regulation and environmental contracts- is in many cases more effective in changing the responsible industry's behaviour than a top-down regulatory enforcement style.[1] Environmental contracts are used as an alternative to action plans, regulations and rules. The concept of 'environmental contract' should not be narrowly interpreted - it is used as a general term for agreed-upon strategies, as well as for legally binding contracts.

The motives of the parties engaged in public-private co-operation to use environmental contracts are diverse.[2] One of the public interests is, adhering to the polluter-pays principle, to distribute among the responsible parties the research, development, and reconstruction costs, etc., which the public sector

---

[1] The bottum-up approach was the basis for the thesis of Keith Hawkins in his book Environmental Enforcement - Regulation and the Social Definition of Pollution, Oxford, Clarendon Press 1984, reprinted 1993, xiii-xiv.

[2] See Ellen Margrethe Basse, The Contract Model, the Merits of a Voluntary Approach, in Environmental Liability Vol. 2, Issue 3, 1994, pp. 74-84; and Lawrence Susskind and Gerard McMahon, The Theory and Practice of negotiated Rulemaking, Yale Journal on Regulation, 1985, pp. 133-165.

P. Glasbergen (ed.), Co-operative Environmental Governance, 179–199.
© 1998 *Kluwer Academic Publishers. Printed in the Netherlands.*

would otherwise defray in connection with the employment of traditional environmental measures. The commercial attractiveness of environmental contracts is that they allow environmental goals and means (reduction of pollution, re-use of waste, etc.) to correspond with both the public requirements and the needs of individual companies or a group of companies. For the parties involved, environmental contracts may also be used as a means to introduce structural changes within the branch of trade. Another reason for private participants to take these contracts seriously is the public relations aspect -pollution prevention pays in the consumer society.

Environmental contracts are new instruments based on voluntary participation and use of cost-benefit calculations. Such contracts may imply an obligation to ensure the realisation of political action plans and a possibility for the organisations involved -or those authorised by the contract- to benefit from economic incentives (for example, subsidies, fees and charges). That is, the contract can call for the use of authority of a public law nature. At the same time, participation in contracts may interfere in free trade. Thus, the contracts may necessitate control of monopolising relations and similar interference in free trade. Monopolies, however, may imply market failures. These, in turn, may imply failure in market forces. Those forces are presumed to offer protection against an ineffective development in situations where regulation not pertaining to public law makes demands on production and production processes. Thus, there is a need for certain political or legal instruments and limitations in the freedom of contract.

In the following, the use of environmental contracts will be examined as a question regarding the connection between trade and environment. This issue is placed in the perspective of EU policy on voluntary environmental agreements.

9.2 EU policy on environmental agreements

The OECD Secretariat, as well as the EU Commission, have taken an active part in the analysis and promotion of such voluntary instruments as part of a more market-based environmental policy.[3] Legal acceptance of the contract as an environmental policy instrument in the EU has been expressed in the Fifth Environmental Action Programme.[4] With the acceptance of new methods for

---

[3] See OECD Environmental Directorate group on Economic and Environmental Policy Integration Voluntary Agreements in Environmental Policies, November 1993; and the Communication from the Commission on Environmental Agreements, COM(96)561 final of 26 November 1996.
[4] Council Resolution 93/C 138/01, O.J.C 138, 1 May 1993, pp 28ff.

standardisation in the EU, a decision has been made to use voluntary strategies (such as environmental contracts) to implement the essential requirements', which can be applied to various products. On 27 November 1996, the Commission published a Communication to the Council and the European Parliament on Environmental Agreements[5], stressing that the success of these instruments requires quantified objectives to be set, the cost of the achievement of such objectives to be moderate, and the number of participants to be reasonable. It is also stressed that certain risks must be avoided - in particular, the setting of vague objectives, lack of transparency, and the impossibility of enforcing the norms included in the contracts. Also the problem of 'free riders' (i.e., non-signatories to the contract) is identified as a risk which has to be avoided. Under some conditions, Member States -or public parties- are allowed to make contracts on national or local procedural rules, quality standards and quota as a means to ensure implementation of EU directives. However, they do not have total freedom to 'regulate' by environmental contracts. Contracts must never conflict with the EEC Treaty's rules and regulations about non-discrimination or equality (Article 7), prohibition of quantitative restrictions of import and export (Articles 30, 34 and 36), common market and free trade (Articles 2, 3g, 85, 86 and 90), harmonisation of specifications for products (Articles 100 and 100a)[6] and observance of the polluter-pays principle (Article 130r(2)). The 'polluter-pays' principle has been recognised as a general principle of international environmental law as well as of trade law, which is important as it forms the basis of numerous contracts. In this area, there are several other principles, e.g., the 'principle of economic efficiency of environmental policy' and the 'principle of just distribution of cost reduction of pollution'.[7] The Member States must, furthermore, abide by Article 5 of the Treaty (stipulating the 'principle of loyalty'). There is also a demand for establishing some guidelines concerning national contracts and the use of different principles in order to avoid misinterpretation of the concept.[8] One criterion which should be included in such a guideline is the 'principle of least intervention'. This principle, which could be called a special application of the public law 'principle of proportionality', implies

---

[5] Com (96) 561 final

[6] See e.g. Jan H. Jans, European Environmental Law, Kluwer International Law, 1995, pp. 89-102 and pp. 106-113.

[7] Hans Chr. Bugge, The Principle of 'Polluter-Pays', in economics and law. In Erling Eide and Roger van den Bergh Law and Economics of the Environment, Juridisk Forlag, 1996, pp. 53-90.

[8] According to the Commission, the framework criteria will have to satisfy inter alia the following generalconcerns: (1) the criteria must be compatible with possible constraints found in national constitutional law - and(as far as possible) in harmony with public law, (2) the criteria shall (as far as possible) be in harmony withexisting private contract law concepts and traditions in all Member States, and (3) the criteria shall ensure thecredibility, reliability and transparency of the contracts.

that the States should always consider whether the objectives aimed for could be achieved by way of environmental contracts, before considering the use of a more restrictive instrument.[9] This principle or criterion must be applied in connection with the 'principle of loyalty'.

9.3 Trade conditions

When an environmental contract is used in a Member State, questions may arise about the degree to which the parties establish monopoly-like conditions or take advantage of dominance in the market, contrary to trade rules. Joint-venture systems based on long-term contracts for collection and recycling product-related waste products could be treated as a market-dominating enterprise encompassed by Article 86 of the Treaty.[10] The anti-competitive effect of collective take-back systems must also be scrutinised under Articles 85 and 86, either directly or in conjunction with Articles 3g, 5(2) and 90.[11] According to Article 85(1), environmental contracts must not create competitively distorting relationships. This provision is based on a principle of prohibition of trade-limiting contracts and concerted practices that can influence trade between the States and which intend to limit or distort competition. Both horizontal and vertical contracts are included in the provision,[12] which also includes non-binding contracts such as gentlemen's agreements and covenants. A contract that falls under the prohibition may gain validity if the Commission, in accordance with Article 85(3), exempts the contract. The competitively limiting part of the contract is not valid, as the Commission may force the participating enterprises to bring the observed non-compliance to an end.[13] By notifying the Commission about a contract, as required by Article 85, and possibly by applying for individual exemption in accordance with Subsection 3, the enterprises can protect themselves against being penalised.

For the individual members of a branch organisation, expenses of environmental measures are typically very different because of divergence in the

---

[9] This principle - which is a principle concerning the competence of administrative authorities - should, however, only serve as a guideline for the legislators' reflection on the choice of instrument.

[10] See Eckard Rehbinder, *Take-back Obligations and Commitments in the light of the EC Treaty*, in "Second Nordic Conference on EU Environmental Law", Copenhagen May 1996, p. 24.

[11] See Rehbinder op. cit. p. 17 and pp. 22-26.

[12] Horizontal contracts concern only one stage in the production or marketing of a product, while vertical contracts comprise several stages, for example production, wholesale and retail.

[13] The concept of non-validity is to be understood in agreement with the general national rules of the Member States. Thus, the courts will not validate contracts if parts of their content violate the prohibition of interference with free competition.

actual technologies. One problem connected with entering into environmental contracts in relation to Article 85 can be that uniform, environmental surcharges, collected as part of the contract, could be significant. Those fees might neutralise otherwise different prices covering the costs of the polluter in accordance with the polluter-pays principle. Without such a contract, the enterprises would carry out individual calculations of the expenses associated with the environmental surcharge. By imposing the same environmental surcharges on each product -and thereby removing environmental expenses from the competition- the incentive for the individual enterprise to find a more efficient and more economic production process than that found by competitors is removed. Moreover, the customers will generally consider an environmental surcharge as a charge imposed by the State for financial gain - and not as part of the environmental 'education' of firms and consumers (based on the polluter pays principle). Consequently, environmental contracts may restrain the firms involved from competing with respect to the part of the price reflecting environmental improvements. These considerations were expressed in the Commission's evaluation of the Dutch contract on the storage of chemicals, titled the VOTOB Contract.[14] After making the environmental contract with the Dutch government for the improvement of environmental standards, VOTOB increased their customer prices by fixed and uniform fees. The fees should have covered the necessary investments for reducing air emissions from the consortium's storage tanks. The Commission pointed out the competitive concerns associated with the environmental fees. The idea was then given up by Dutch contractors associated with VOTOB, with the consequence that the Commission discontinued the procedure.

National environmental agreements almost invariably have an adverse impact on imports. Consequently, such agreements conflict with the fundamental market freedoms, especially the prohibition of measures equivalent to quantitative restrictions laid down in Article 30 of the Treaty.[15] To give an example, the take-back obligations laid down in some contracts are an equivalent measure covered by this Article[16] Under the Dassonville formula the Article covers all

---

[14] VOTOB is the abbreviation for the Dutch consortium of companies offering tank storage facilities to third parties in Amsterdam, Dordrecht, and Rotterdam.

[15] See Eckhard Rehbinder Take-back Obligations and Commitments in the light of the EC Treaty in 'Second Nordic Conference on EU Environmental Law', op.cit.

[16] It is stressed by Eckhard Rehbinder (supra note 11), pp. 14-15, who draws the following conclusion: "Foreign producers and importers of foreign products, due to their distance from the domestic market and lower sales on that market, will often have more difficulties in establishing a comprehensive system of take-back, transport to, and recycling in their home country, especially where the relevant member state law extends the take-back and recycling obligation to products of the same kind sold by competitors (which might be the case in relation to products with standardised packaging). On the other hand, the possibility of charging enterprises located on the import market

measures which are capable of hindering -directly or indirectly, actually or potentially- intra-Community trade. If the Member State, however, can prove that a national environmental contract is necessary in order to satisfy mandatory requirements relating to the protection of the environment, it will be justified on the basis of the Cassis de Dijon Principle.[17] The European Court of Justice may, however, review national equivalent measures laid down in contracts as to the existence of less burdensome alternatives as well as to the question whether the costs incurred by foreign producers are out of proportion with the benefits derived from the protection of the environment in the Member State. It is the State that has to prove proportionality and justification of the relevant conditions.[18]

In order to be in harmony with EC trade law, a restrictive measure laid down in an enviromnental contract must represent the means which restrict the free movement of goods in the least disturbing way.[19] This limitation of national regulation was stressed in the famous Danish Bottles Judgement by the European Court of Justice (ECJ).[20] The court case concerned a Danish statutory order which prescribed an obligatory take-back system with respect to soft-drink and beerbottles.[21] The business concept in relation to the design of bottles had to be approved by the Danish Environmental Protection Agency (EPA). It was clear that approval would be refused if, in particular, the bottles in question were not technically suitable, the take-back scheme did not ensure a sufficiently high rate of re-use, or if an approved bottle of equal capacity was already available. The reason for the requirement of approval was that any type of authorised bottles could be returned to any shop selling beer and soft drinks, thereby achieving as high a return rate as possible. It was maintained by the Danish authorities that in order to make this comprehensive system workable, the maximum number of bottles and take-back schemes approved at one time had to be limited to around thirty. Some modification to the last-mentioned requirements was hereafter laid

---

with compliance with the take-back and recycling obligations will normally have a cost equalising result and remove the competitive disadvantages foreign producers and importers have vis-à-vis their domestic competitors. However, where foreign producers primarily use packaging materials that are less easily recyclable than the materials used by domestic producers, disadvantages remain."

[17] This principle was used in the Danish Bottles Case mentioned below.

[18] See Rehbinder op. Cit. P. 12

[19] See the Cassis de Dijon Doctine stressed in case 120/78 1979 ECR 649. The ECJ ruled that - in the absence of EC rules relating to the marketing of products, obstacles to the free movement of goods that arise from disparities between the national rules of the Member States have to be accepted where they are applied in a non-discriminatory manner to domestic and imported products and are applied in order to satisfy mandatoryrequirements recognised by EC law. The ECJ considered as 'mandatory requirements' the necessity for fiscal control, fair trading practices and consumer protection.

[20] The Commission v. Denmark, Case 302/86, (1988) ECR4607, see Rehbinder op. cit. p. 9.

[21] Statutory Order No. 397 of 2 July 1981.

down in a new statutory order (to secure a better harmony with EC rules).[22] This Order accepted sale of a limited quantity[23] of drink in non-approved bottles (except metal containers). Sales to test the market were also allowed. Deposit and take-back schemes were to be set up for each non-approved bottle - the deposit being no less than the amount generally charged under approved schemes. In effect, under the derogation schemes, bottles could only be returned to the shop at which they were purchased or to another shop selling the same bottles. The result was that a lower proportion of bottles than the approved schemes required were likely to be returned. The Commission argued that the requirement was incompatible with Article 30 of the Treaty. The ECJ held, however, that Denmark was allowed to introduce such a system because it aimed at the protection of the environment. The Court found the requirement for a take-back scheme to be proportionate to the aim of protecting the environment through the re-use of containers: "it must be observed that this requirement is an indispensable element of a system intended to ensure the re-use of containers and therefore appears to be necessary to achieve the aims pursued by the contested rules. That being so, the restrictions which it imposes on the free movement of goods cannot be regarded as disproportionate."[24]

The statutory demand for approval, however, was disproportionate to the environmental aim and put an excessive burden on intra-Community trade. The extension of give-back opportunities offered at sales outlets that sold other products in the same kind of bottles, in the view of ECJ, would amount only to a limited improvement of the recycling rate, an impact which was out of proportion with the burden on trade. The new Danish incentives within this area will be described below.

The question of subsidies
The question of *subsidies* is one of the most important in the future of the trade-environment relation and the compatibility of use of environmental contracts.[25] If the authorities in EU Member States enter as parties to an environmental contract, questions may arise about the degree to which the contract provisions imply State aid for domestic firms and conflict with the

---

[22] See Statutory Order No. 95 of 16 March 1984, which was initiated in response to the criticism which the Commission laid down in a letter of 21 December 1982.
[23] Max. 3000 hl each year of every producer.
[24] See point 13 of the judgement.
[25] See Benedict Kingsbury, 'Environment and Trade', in A. Boyle (ed.) Environmental Regulation and EconomicGrowth, Clarendon Press, Oxford 1994, pp. 214-216, concerning the rules on subsidies in the GATT/WTO Agreements.

polluter-pays principle.[26] According to Articles 92-94 of the Treaty,[27] it is prohibited to give direct or indirect public aid to a group of firms or products if that aid influences the competitive situation in the Common Market. In addition to the literal interpretation, the term 'State aid' is also understood as being aid from regional and local authorities. In cases of doubt, the schemes that could be construed as State aid must be notified. Notification is a condition for validity -a scheme cannot legally be put into effect at national level before the Commission has made its final decision. The prohibition on State aid includes not only direct subsidies but also indirect aid - for example, loans on favourable terms (e.g., special loan guarantees), special tax conditions, favourable price setting, reimbursement for collection regimes, etc. (see above concerning the VOTOB case). In EU practice, national environmental subsidies have been tolerated pursuant to the guidelines laid down in the 'Community Frameworks for State Environmental Subsidies'.[28] Also, environmental contracts have been accepted as being based more on environmental conditions than on state aid conditions. This will be illustrated below through information on the Danish Used Tires Contract.

9.4 Public or private law contracts

The most complicated and important type of environmental contract at the national level is a contract between the State and one or more organisations on behalf of their member enterprises. By employing such contracts, organisations or enterprises may consider using particular instruments for environmental improvement (such as production modification, changed waste disposal, etc.). The State may promise to implement (or not to implement) certain international, EC or public law obligations. In situations where the State is a party to the co-operation, such contracts, which are primarily characterised as 'gentlemen's agreements', or convenants, must be categorised as public law contracts because they involve the removal of common environment nuisances and dangers and involve or limit the use of traditional public law instruments. With respect to regulated firms, the scope of the contracts will typically be supplemented by the parties to the contract, including the branch organisations, providing flexible

---

[26] In any case, the basic point is that if the public sector acts as any other rational investor would act under ordinary market conditions, economic incentives (such as that for the recovery of lead storage batteries) are not State aid, in the sense given by the TEU.

[27] The public sector is prohibited from giving: ". . . any aid granted by a Member State or through State resources in any form whatsoever which distorts or threatens to distort competition by favouring certain undertakings or the production of certain goods... in so far as it affects trade between Member States."

[28] The Commission's guidelines date back to November 1974 and were amended in March 1994.

conditions for the 'regulated' enterprises (for example, information, branch guidelines, quota, etc.). The branch organisations engage in ensuring that the organisations' members handle, collect and dispose of waste, etc. by the agreed-upon methods. Further, the organisations can, through the contracts, undertake to establish administrative functions and to ensure that fees/environmental surcharges, etc. are paid or collected.

Public law aspects are expressed very clearly in the supplementary 'command and control' competence, which the Minister is given by a rule in the Danish Environmental Protection Act (see below).[29] The intention is to ensure that non-parties to the contract become regulated through rules laid down in statutory orders (anti-free-rider rules). Therefore, there is a reason for ascribing public law considerations a special weight in evaluating such contracts. At the same time, public law considerations must be used with the understanding that it is not first and foremost the individual's demand for law and order which needs to be ensured in this situation. Rather, the common interest of different groups must be safeguarded. Thus, the principles of administrative law must be conscientiously used. Such considerations are also required because of environmental principles (codified in the acts, in the TEU and in the international conventions) which can be in direct conflict with the principles of administrative law. This applies not least to the precautionary principle, which -in contrast to, for example, the principle of proportionality- gives the environment the benefit of the doubt, while the principle of proportionality gives the addressee (the polluter) the benefit of the doubt. The private law relationship between organisations and their members is rather complicated in the situations where the organisations enter into contracts with the authorities. The rules of the organisation, which the members have accepted, become very significant for the organisation's capacity to commit its own members to a contract. The prerequisites for members joining the organisation which are expressed in the rules cannot be shifted without the members having had the opportunity to reconsider the basis on which they originally joined the organisation. Neither can economic obligations arising from entering into a contract be imposed on an organisation's members if they are out of proportion with the obligations inferred from the original charter.

Some environmental contracts primarily fall within the field of private law because they involve only the regulation of relationships between private parties. Such private law contracts are, for example, contracts between branch organisations or between enterprises. As long as the contract is between private parties, its conditions are primarily regulated by contract law. Such contracts become significant also for non-parties - e.g., for consumers who have to pay for

---

[29] Act No. 358 of 6 June 1991 - which came into force on 1 January 1992. The rule mentioned in the text is found in Section 10(4).

the environmental measures implied by the contract. Transferring environmental expenditure to the final consumer through price increases can be totally in compliance with the polluter-pays principle. The contract may, however, comprise environmental damages of both the past and the future. In such cases, the branch seeks a solution through which the current consumers pay for earlier consumers' and producers' lack of environmental consciousness.

A typical private law contract concerns the exchange of services between parties. Lack of consensus in making the contract will mean that the exchange (which according to the assumptions of the contract should be advantageous for both parties) cannot be realised. But lack of consensus with respect to a contract within a branch does not necessarily mean that the future environmental requirements on that branch will be changed. Environmental authorities follow the course of negotiations and the subsequent results with interest. Environmental policy will be drawn up and implemented whether or not agreement exists among the parties. Accordingly, making contracts has no decisive significance for the substance of the regulation.   Private law contracts will often be supplemented with public law contracts or statutory rules to ensure both efficiency and appropriate advantages for the regulated parties. So, contracts between organisations and/or members can be supplementary to an environmental contract regulated by public law entered into with the State, for the purpose of remedying the limits of authority which, according to association law, characterise the situation of organisations today. However, private law contracts, in which only private parties participate, can also stand alone.

9.5 Danish law and contracts

A characteristic of environmental legislation in Denmark has been the lack of a true multimedia approach. Instead, a complicated administrative system, in which several decision-makers (private as well as public) need to but will not always collaborate has been created. As a tool to change the public law tradition, the revised version of the Danish Environmental Protection Act from 1992 (hereafter, the Act), contained a rule in Section 10 providing for the use of environmental contracts - and, consequently, accepting public-private co-operation as part of the regulation style. According to Section 10(1), the Minister for Environment and Energy (hereafter, the Minister) has the authority to establish a binding set of objectives with respect to the environment. According to Subsection 2 of the same Section, the Minister can make environmental contracts with firms or branch organisations. In examples of environmental contracts currently in force, branch organisations are normally parties to the contracts.

To ensure equality and that contracts are fair to all parties involved, firms must be inhibited from behaving as 'free-riders'. According to Section 10(4), the Minister has the power to compel administratively relevant parties that have not agreed to an on-going contract to meet the same environmental obligations as laid down in the deal. The Minister can, accordingly, in the form of a statutory order applying to non-signatories: "...establish corresponding demands for limiting the use, discharge or disposal of products, chemicals, other materials..." [author's translation].

In the above context, corresponding demands' are understood as demands aiming at a reduction, etc., equal to that established in the contract. However, in this connection, differences shall be accepted. That is, although firms are merged for the purpose of achieving a collective reduction goal, the firms will not be held equally responsible if they damage the environment differently.

Pursuant to Section 10(5), the Minister can determine that decisions can be changed in concordance with on-going contracts with respect to the special rules issued under Subsection 4 of the Section.

In 1993, Section 9a was added to the Act to allow for schemes whereby producers could recover used products and establish flexible collection and processing systems. Reference to the use of take-back contracts to establish such schemes is only made in the explanatory statements to the Act. By accepting these explanatory statements, and as a result of several debates on the choice between the use of contracts versus the use of rules, the Danish Parliament has accepted the use of take-back contracts as an alternative to traditional norms and decisions.

Take-back contracts are not legally binding to the same extent as environmental contracts entered into in accordance with Section 10 of the Act. The explanatory statements to Section 9a do not suggest any possibility for properly sanctioning take-back contracts, nor does it give the Minister power to supplement binding rules for non-parties to the contract. The only sanction applicable for non-compliance with the contract is the Minister's use of authority to issue binding rules in a statutory order or Parliament's use of legislative authority.

In the following, some of the Danish environmental contracts will be discussed briefly and elucidated from the perspective of the legal regulation of trade and environment that has been decisive for the formulation of the contracts. The contracts reflect the character of the actual environmental problems which they aim to resolve. Most of the contracts have been influenced significantly by the review made by the EU Commission and national competition authorities, a process which has been used to determine which contracts and/or elements of contracts are considered valid.

The lead storage battery contracts.

In September 1993, ReturBat, the association collecting lead storage batteries in Denmark, and the Association of Danish Recycling Industries entered into a private law contract (not covered by Section 10 of the Act) for the collection and re-use of discarded lead storage batteries. The aim of this contract was to ensure an environmentally defensible collection and re-use of discarded lead storage batteries in order to avoid inappropriate disposal, including uncontrolled landfill and incineration activities. This aim was in harmony with the international obligations laid down in the EC Waste Directives[30] and the Directive 91/157/EEC on Batteries and Accumulators. The objective of the contract was to ensure a recovery rate greater than 95 per cent.

In the contract, ReturBat committed itself to secure the necessary means for a controlled and environmentally correct collection and re-use of waste materials. The Association of Danish Recycling Industries had to make sure that its members were qualified to collect all the lead storage batteries discarded in Denmark and that they -in consideration of the resource utilisation of discarded lead storage batteries- stored and transported these batteries in compliance with the applicable environmental regulations.

In May 1995, this contract was supplemented with another environmental contract negotiated under public law by ReturBat and the Minister. The contract covered all forms of rechargeable lead storage batteries sold in Denmark (sealed as well as non-sealed types). The objective was to reach at least a 99.9 per cent collection rate by the year 2000. ReturBat was engaged to ensure that all information necessary for the yearly calculations of the collection rate was gathered. The contract included a fee imposed on all lead storage batteries, the size of which was to be established by the Minister through a statutory order. The contract was supplemented with drafts of two statutory orders: one concerning the fee on lead storage batteries; the other concerning the collection of lead storage batteries, the subsidy for commercial collection, and the removal for re-use. The statutes were brought into force as the Statutory Orders No. 91 and 92 of 22 February 1996. The element of voluntariness in the construction of the contract is, thus, clearly supplemented by traditional legal norms. At the same time, the regulation pertaining to public law created an economic basis for the scheme by introducing fines used to finance the collection.

From April to June 1995, the Danish Competition Council evaluated the public law lead storage battery contract. From the start, the Council was critical about the drafts presented. It was the Council's opinion that all firms qualified to ensure an environmentally correct disposal ought to have the same opportunity for

---

[30] See as the most important ones the Directive 75/442/EEC on Waste - as amended by 91/156/EEC - and the Directive on Dangerous Waste 91/689/EEC, as amended by 94/3 I /EC.

receiving part of the subsidy, which, according to the draft, could only be paid to ReturBat. The Council was also critical of the exclusive right held by ReturBat's members, given the original presumption that firms should be able to collect nation-wide, etc. The demand for a national coverage, directed towards the firms who wished to participate in the contract, implied a risk of monopoly. Consequently, the contract was changed, enabling the participating parties - and hereby potentially subsidised firms- to at least carry out county-wide collection. Regarding the designation of competence, the Council was concerned about ReturBat administering the collection scheme, mostly because there were no contractual mechanisms for appealing decisions made by ReturBat to an administrative authority. The construction implied a risk that the branch organisations would safeguard the interests of their own members - and thereby monopolise the market. The Council recommended that the contract be reformulated to clearly state that a collection firm could receive its subsidy if it could prove that it would send the collected lead storage batteries on to a firm that was environmentally approved to handle lead storage batteries in a defensible way. The result was, thus, that the use of environmental contracts was accepted by way of a concrete weighing of the environment and trade freedom. The weighing, however, justified considerable changes that were accepted by the parties to the contract.

Also, the Parliament and the Ministry for the Environment have clearly indicated how far they want to go in acceptance of a voluntary solution. If the contract's objectives were not realised, the Minister should take the initiative to implement, via rules, a more adequate scheme that would improve the possibility of achieving the environmental goals. The contract and the accompanying statutory orders must, therefore, be seen in connection with the two acts passed in May 1995: Act No. 414 of 14 June 1995 concerning a Fee on Lead Storage Batteries and Hermetically Sealed NiCd Batteries and Act No. 404 of 14 June 1995 on Remuneration in connection with the Collection of Lead Storage Batteries and Hermetically Sealed NiCd Batteries. The acts will be able to substitute for the voluntary contracts if they fail to be effective enough. Neither the Acts nor the contract must violate Articles 92-94 of the Treaty or the polluter-pays principle. In December 1995, these Acts were reviewed and accepted by the Commission in accordance with Article 93. After a successful negotiation with the branch on the content of the contract, the Minister in March 1996 brought a bill to the Danish Parliament suggesting an amendment.
Consequently, Act No. 404 and 414 do not cover these batteries; the acts only cover the materials sheltered by another contract - the NiCd Battery contract. An assessment of effectiveness as well as competition thus prompted an acceptance of a voluntary scheme for the lead storage battery waste problem.

The VOC contract

In 1994, an environmental contract for reducing the emissions of Volatile Organic Compounds (VOC) was made between the Confederation of Danish Industries and the Minister. In co-operation with the Confederation, the Minister took on the preparation of branch guidelines for the branches participating in the contract.

The guidelines associated with this contract were presumed by the EPA to be used by the supervisory authorities with respect to all firms within the relevant branches, including the branches and firms that are not parties to the contract. The character of the contract is primarily that of an action plan for the branches involved. It is based on technical statements about the possibilities of emission reduction as well as statements about the possible technical and economic barriers to reduction. For the industry the advantage of this contract lies precisely in the contract itself. The EPA promised, already in 1989, not to tighten the threshold limits for VOC in air quality guidelines if the industry would prepare a plan for substantial reduction of VOC emissions by the year 2000. The possibilities of sanctions were not included in the contract. The Competition Council examined the contract in September 1994, as it wished to ensure that differential treatment would not result in monopolising relationships. The Council was most concerned about the following provisions in the contract: "The contract committee ... will work together to make sure that the contract is respected by the central authorities." [author's translation] and "The committee of supervisors has the responsibility of informing local environmental authorities about which firms are included in the contract." [author's translation]. With such terms of contract, the authorities take part in a structural change within the branch of trade, which may imply a monopolistic position for the parties to the contract.

According to information from the Confederation to the Council, the contract did not preclude organisations and firms that were not members of the Confederation from subsequently being covered by the contract on the same conditions as the branches and firms which were covered by the contract as members of the Confederation. At the same time, the Council declared that the provision about passing on information to the local authorities about participation in the contract was unnecessary. The reason was that the EPA had declared that the local environmental authorities would not let the question about contract relationships affect the decision-making in cases on environmental approval of the firms specified in the Act.

After the EPA had informed the Competition Council that preferential treatment was not intended, and after it was disclosed that the possibilities of being included in the contract were open, automatically conferring the related advantages to those firms joining, the Council declared the contract not subject to the obligation to notify. After the changes proposed by the Competition Council had been effected by the contracting parties, the Council accepted the contract. This case illustrates that the participation of the authorities in a contract, that may imply a monopolising position, will not be accepted. Also in several other cases, the Danish Competition Councils and/or the Commission have assessed the participation of the authorities. As an illustration, see the details on the Used Tires Contract below.

The bottle contract

The Danish interest in the re-use of bottles is very strong. The Minister for the Environment made a contract in March 1991 and another one in November 1994 with the Danish soft-drink producers, Danish retailers and the Danish Consumer Council. The first contract authorised the recycling of 1½ litre soft-drink bottles, the second pertains to ½ litre bottles. The contract only applies to Danish producers. The authorisation, which was based on the authority given in the Statutory Order on Packaging of Beer and Soft Drinks, was part of the Minister's obligation -or acceptance of the EPA's decision- stated in the contract. The producers of soft-drink bottles were under an obligation to jointly develop standardised bottles, to no longer import bottle variations, and to take some glass bottles off the market. The retailers were, as part of the contract, to accept the take-back of all bottles that they themselves distributed. Obligations concerning cleaning, the use of uniform cases for bottles, the sorting of returned bottles, as well as a co-operative development for making the take-back of bottles easy also became part of the contract.

As a part of the contract, the EPA agreed, in co-operation with the private parties, to undertake an analysis of the public regulation -and the need for regulation- in respect to bottles for beer and soft drinks with regard to securing maintenance of an effective take-back system. The EPA agreed to assess the development in the total take-back percentage, including that of the new PET bottles. If an effective system is not achieved, the contract will be revised among the concerned parties.

In November 1994, in connection with the contract, the Minister asked the EPA to announce to the producers, as well as to the retailers and the Consumer Council, that the need for different beer and soft-drink bottles in the Danish take-back bottle system had been fulfilled. The announcement was based on the interest in ensuring an effective take-back system in which nearly 100 per cent of the bottles could be returned for refilling. In the future, given this announcement, only standardised bottles that can be used by all parties in the industry will be accepted.

These contracts are examples of how environmental contracts can be used as an integrated part of normative regulation. The contracts are not an alternative to decisions or rules. They are only meant to ensure an administration that is restrictive and which the regulated business as well as the public would consider to be observing the principle of fairness. As the contracts only apply to Danish producers, the EU's interest with regard to the aspects of competition is limited. However, it could to some extent be alleged that the take-back system violates the Treaty's Article 30 and 34. As mentioned above, environmental concerns may form the basis of certain import and export restrictions. The difficulties with which importers may faced regarding take-back obligations is described by Eckard Rehbinder in his article Take-back Obligations and Comments in the Light of EC Treaty.[31] The subject shall not be elaborated here.

At the moment, the Danish bottle rules and the international obligations laid down in the trade rules and in the Packaging and Packaging Waste Directive (94/62/EC) are being discussed.[32] In June 1997, Denmark received an Eu statement of objections from the Commission because of a general prohibition of metal containers. The Directive has an important impact on the production, marketing, and distribution of packaging products. It is the opinion of the Commission that the Danish rules conflict with these obligations. The national schemes -including environmental contracts- for achieving the target for recovery of packaging waste shall not create any trade barriers, or distort or reduce competition in the market. The ECJ Judgement on the Danish Bottles case does not include an evaluation of the Danish prohibition of use of metal containers. Therefore, the judgement cannot be expected to support any arguments pro or contra the Danish prohibition of use of metal containers (e.g., the Swedish re-useable containers). The environmental contract is not included in the case; thus, the weighing of the environment and trade freedom, expressed in the contract, is not under review at the moment. With regard to Denmark, however, it should be mentioned that the environmental contract excludes new producers from having any influence on the types of containers that may be used and puts them in a worse position than foreign producers.

The contract on used tires

Upon passing the motion for a decision (B 31) on increased recycling, the Danish Parliament prevailed upon the Government to see that a system for the recovery of tires was established. The interest for making this a reality was further expressed in the 'Danish Action Plan for Waste and Recycling 1993-1997'. After several years of negotiations, a contract on the collection and re-use of tires from motorcycles, private cars, commercial vehicles, etc. was finally, in February 1995, signed by the Minister, representatives of the tires and automobile organisations as well as the municipalities. The contract only affects tires collected after 1 July 1995.

---

[31] See Eckard Rehbinder, supra note 11.
[32] See O.J.L 365 of 31 December 1994.

The target was to increase the percentage of tires recycled within a five-year period from the current 35 percent to 80 percent. The purpose of the contract is to avoid the deposition of tires to landfills as well as to ensure a high utilisation of the materials, making recycling a priority over energy utilisation. These goals are in harmony with the EC waste law and strategy, especially the new Strategy from February 1997. Included in the last waste management strategy and in the new Directive on Waste Disposal is a ban on landfilling with tires, alongside a new focus on materials recycling and incineration with energy recovery.

Firms selling tires are obliged by the Danish contract, without extra charge, to take back as many discarded tires from the consumers as they have sold. Accordingly, the firms entering the scheme have agreed to collect all used tires -including tires which the consumers deliver to the municipalities- as well as to make sure that the required capacity exists for re-gumming, materials re-use and incineration, as expressed in the contract. The Commission was informed by the Danish EPA that approximately 50 firms were expected to participate in the scheme and that the possibility was open to other firms -including firms from other Member States- to participate in collection as well as delivery to special facilities in Denmark. In order to participate in the contract, a firm must be approved by the EPA and be able to document the environmental approval of the activity given according to the Act.

Like most contracts, this one has an 'economic basis' in legal rules pertaining to public law, ensuring that the consumers pay for the services of the branch of trade. The fee scheme -which became effective on 15 March 1995- was established through Statutory Order No. 144 of 3 March 1995, which introduced both the fee and a subsidy for the disposal of tires. According to the Order, upon purchase in Denmark, a fee of DKK 8 is paid for new tires and DKK 4 for used tires, thereby creating an economic incentive to buy used tires. The fee is not imposed on tires that are exported, as the disposal of such tires does not contribute to the environmental problems in Denmark.

The contract and the related statutory orders were examined by the Commission, according to Article 92 of the Treaty. The contract, including the provisions about a subsidy, was accepted because the services (which the participating firms had agreed to undertake on a voluntary basis) were compensated or paid for if the tasks were carried out for the benefit of the general public interest. The Commission added that the system was in agreement with the polluter-pays principle and that the consumer -in accordance with this principle- in reality would pay more of the environmental costs under the agreed scheme than under the general (currently valid) waste legislation/schemes, in which the tires would only be landfilled. At the same time, the contract would reduce the risk that used tires would be exported to other countries, as none of the established schemes gave an economic incentive for such export. The Commission supported the idea that, in accordance with the contract, payment for collection would be made conditional on delivery of the used tires to a rubber-reducing facility that was environmentally approved for this activity. It was essential that the price of used tires to be paid upon delivery to approved processing facilities was agreed on without interference from the Danish Govemment. It was also emphasised that the price for treatment was lower than the price would otherwise be -public and private costs, etc.- in connection with disposal of such waste. On this basis, the Commission concluded that the environmental contract did not violate Article 92. Thus, the contract was a manifestation of an acceptable weighing of interests between the environmental concems and trade freedom.

9.6 Trade-related evaluation criteria

This chapter seeks to give some guidelines for answering the questions: is there a harmonious relationship between the use of environmental contracts and the duty to comply with the trade policy? And what is the relation between the use of contracts and the condition in environmental and trade-related policy that the polluter shall pay for pollution prevention and re-establishment? Compatibility between an instrument and principles is difficult to demonstrate. The response to the question depends on the characteristic elements of the instruments and on the content of the principles. But it also depends on a very complex interaction between environmental law, competition law and the development of technology as well as the economy. The above examples of assessment of environmental contracts -in particular the contract on used tires- suggest that in practice it includes an assessment of the environmental aims and efficiency of the contract and the polluter-pays principle.

Contracts which imply the *use of State aid*, environmental fees, take-back schemes, fixed fees, etc., all of which could reduce the firm's incentive to invest in cleaner technology, from an economic/market viewpoint, could become invalidated. This is what happened after the Commission's review of the Dutch VOTOB Contract and the Danish oil Industry's Environmental Contract, which is not included in this chapter.[33] The review of the Contract on Used Tires stressed that if such a scheme did not include the State as a party to the contract, if the fixed environmental fees, etc. were gauged and used in a way that secured the observance of the polluter-pays principle in connection with disposal, and if

---

[33] See concerning this Ellen Margrethe Basse 'Environmental Contracts - as Tools to Implement International Obligations', in Ellen Margrethe Basse (ed.) Environmental Law - From International to National Law, GadJura, 1997, pp. 151-153.

the export of waste for disposal was reduced, then the contract could be accepted, despite the prohibition on State aid.

The experience with contracts makes it possible to list some other criteria that can be used to evaluate the future potential of environmental contracts at the national level from the perspective of trade law - and by doing so also in some relations evaluate the compatibility with the polluter-pays principle. The criteria, which are used by the authorities to evaluate competitive effects, are summarised here and used to reveal the potential limits to entering into environmental contracts.

The possibility that *exclusive contracts* would not be accepted is expressed in the Danish Competition Council's evaluation of several contracts: e.g., the Lead Storage Battery Contract, and the VOC Contract. The Council presumed that all firms that could carry out an environmentally correct disposal, as evaluated by objective criteria, ought to have the same opportunity of participating in the collection scheme for lead storage batteries and of receiving part of the subsidy. The parties in the VOC Contract were also forced to remove elements that might work to distort trade. It was explicitly stressed by the Confederation of Danish Industries that the contract should be taken to mean that no firm would be held outside the contract because of its non-membership of the association. In the evaluation of a contract made by the Refrigeration Industry, the Council emphasised that the approval procedure and the assumed boycott of non-approved enterprises was an expression of an unfounded exclusivity, which could not be accepted.

In assessing the contracts in the Commission, Article 85 will lead to application of similar criteria. The lack of examples in this connection may be explained by the fact that most Danish contracts - though this is presumed - are not notified to the Commission.

The problems of incapacity were relevant in the Danish Contract on Lead Storage Batteries. The Competition Council declared in the contract mentioned above that members of ReturBat should not benefit from the administration plan of the contract when non-members are precluded from benefiting from it. Also, the general principles of impartiality and fairness in administrative law have frequently been stressed as relevant in the evaluation of environmental contracts. How far the Commission is prepared to go into such a more administrative law assessment is not under discussion here. Since the requirements, however, are closely connected to the assessments of trade freedom, it may be expected that the Commission may also use such criteria.

In practice, the demands for participants to cover *nation-wide contracts* have been problematic, as they may exclude other enterprises from the market. The Competition Council found the Lead Storage Battery Contract problematic because of a presumption that a nation-wide scheme could preclude enterprises other than members of the ReturBat from participating in the market. The result of this criticism was that participation in the collection scheme no longer presumed that participants had the capacity -or the obligation- to be nation-wide. Only a regional coverage was assumed. The conditions set by the Commission typically include the ensurance of schemes covering the State in question. This condition is based on the requirement for implementation of EU rules to apply to the whole country in the Member State. The capacity of private persons is not considered in this respect.

When evaluating the content of contracts, it is also important to ensure that the environmental contracts do *not violate the fundamental environmental principles*, but support the policy guidelines included in the principles. As stressed above the polluter-pays principle was found to be relevant in the assessment of the Used Tires Contract.

In reviewing the Lead Storage Battery Contract, the Danish Competition Council emphasised that the contract should be seen in *the light of the principle of proportionality*. That is, the contract should be submitted to a review by administrative law, competition law and EC law to see whether the requirements of the parties to the contract were more extensive than called for by the condition of the environment. If the actual purpose of the contract was to ensure an enforcement of structural changes within the industry, extra strict requirements would be placed on the parties to document that the environmental considerations were so substantial and widespread that the contract should be accepted. This was also an issue in the evaluation of the VOC Contract. With respect to the VOC Contract, the Competition Council found that the draft of the contract contained information requirements that were too extensive.

As already stressed above, some contracts might give rise to problems because of the *presumptions about import and export prohibitions* (laid down in Articles 30 and 34 of the Treaty) and the adequate use of the principles of self-sufficiency and proximity.[34] To the extent that the prohibition of import and export concerns waste disposal in landfills, the contract, which limits transport subject to the claims in the waste regulation with reference to the above-mentioned principles, will be in agreement with the OECD's and the EC's Waste Regulation. This is emphasised in the Commission's review of the Contract on Used Tires, in which it is stressed that contracts which limit export of 'waste for disposal' are in agreement with the EU environmental policy. In a contract about the re-utilisation

---

[34] See concerning the judgement of ECJ in the Wallonian Case (C-2/90, ECR 1992 443 1) Philippe Sands, 'International Law in the Field of Sustainable Development: Emerging Legal Principles', Winfried Lang (ed.), Sustainable Development and International Law, Graham & Trotman/Martinus Nijhoff, 1995, pp. 56-57.

of waste, however, adequate use of the principles of self-sufficiency and proximity cannot justify import and export restrictions.

To close on Denmark: new environmental agreements cannot be expected at the moment. According to a discussion paper from July 1997 on 'Industrial Refuse and Chosen Flows of Waste', the Ministry for the Environment has not been satisfied with the instrument. The Ministry prefers a traditional regulation with focus on producer liability. The Ministry finds that there is no need for initialisation of new initiatives: before an evaluation of the concept of the contract and the privatisation of the responsibility for waste disposal has been carried out, which will be a consequence of the implementation of producer liability. The Danish rules issued in 1997 in the form of new orders on waste increases the monopoly of the municipalities. The monopoly may cover both industrial waste and other kinds of waste. No distraction is made between reusable waste and waste for disposal. The message from the EU that trade freedom shall be drawn into waste regulation has, thus, not been recognised in Denmark. The contracts within the area of waste could have been a means to ensure an appropriate weighing of environmental concerns and trade freedom. In Denmark, however, this line no longer seems to be supported by the Minister for the Environment nor by the Ministry.

10. DEMOCRACY AND ENVIRONMENTAL AGREEMENTS[1]

MARTIN ENEVOLDSEN

10.1 Introduction

Co-operative management denotes a new approach to environmental governance in which public and private parties co-operate on relatively equal terms with a view to reaching consensual and practicable solutions to environmental problems (cf. the introductory chapter to this book). A major policy instrument within this new approach is the environmental agreement. Environmental agreements imply a different procedural method for reaching decisions than direct regulations: the private parties who are supposed to carry out pollution abatement activities participate directly in all phases of the decision process, that is, in preparation, negotiation, formulation, implementation and evaluation of agreements. Thus, agreements are based on interactive processes of environmental policy-making. In principle none of the parties can be forced into an agreement; decisions are only made when discussions lead to a common solution to which all parties can agree. Hence, certain political actors have advanced a democratic image of co-operative management and environmental agreements connoting implicitly to modern ideas about consensual decision-making, discursive democracy and communicative rationality.

On the other hand, environmental agreements have been criticised for their tendency to closed, exclusive decision processes in which only the government and business interests (industrial interest organisations and firms) are allowed to participate.[2] The frequent exclusion of the parliament, environmental groups, and trade unions from co-operative management processes is held to be a serious

---

[1] The empirical evidence for this article is furnished through an ongoing cross-national project 'Joint environmental policy-making in the EU and selected member states' in which I participate along with Duncan Liefferink and Arthur Mol (Wageningen Agricultural University), Volkmar Lauber and Karin Hofer (Salzburg University), and Mikael Skou Andersen (Aarhus University). The project is financed by the European Commission's SEER programme: 'Environment and Climate 1994-1998'.

[2] For some general points in relation to the democratic objections to co-operative environmental management see Meadowcroft, ch. 2 this book. As regard the criticism raised by Danish and Dutch authors concerning the democratic deficits of environmental agreements, see e.g. Bjerregaard (1995) (Environmental Commissioner). Kjellerup (1995), Willems (1990), Van Buuren (1993), Biekart (1995).

P. Glasbergen (ed.), Co-operative Environmental Governance, 201–226.
© 1998 *Kluwer Academic Publishers. Printed in the Netherlands.*

democratic problem which undermines the legitimacy of environmental agreements. The criticism along these lines stress equal rights of participation thereby connoting to the fundamental values of liberal democracy.

Several questions arise from this brief presentation of the problem. First of all: do environmental agreements in practice entail a democratic paradox, promoting some aspects of democracy and undermining others? Second: are environmental agreements really superior with respect to communicative qualities of democracy and deficient with respect to liberal virtues of democracy such as equality, liberty and fairness. Third: on balance, are environmental agreements more or less democratic than traditional forms of environmental governance? To my knowledge there exist no theoretically informed analysis of these and related problems concerning the democratic qualities of environmental agreements. What follows below is a first attempt to answer at least the first two of those three questions.

In my point of view the problems cannot be answered by building further upon the common sense arguments on the processual advantages and disadvantages of environmental agreements. Nor do I think it is wise to superimpose a set of more or less arbitrary democratic criteria derived from the varieties of current political practices. Such a strategy would never free itself from the functionalist reasoning that environmental agreements are just a new way of understanding democracy, refuting any claims of a democratic deficit. In contrast I will follow a more critical, normative approach which assumes that, after all, democracy has a more fundamental meaning to us, rooted in historical occidental values. This implies that the democratic evaluation of environmental agreements must be preceded by a clarification of these values and an elaboration of the corresponding democratic norms.

I will pursue this task in the first half of the paper by arguing that two alternative democratic perspectives -one called fairness of participation, the other communicative qualities of participation- are equally relevant in assessing the democratic qualities of any political process. The norms advanced through the perspective 'fair participation' are inspired by the liberal democratic values of liberty and equality as they are interpreted by John Rawls and Robert Dahl. The other perspective, which relates more to values associated with the community, will be elaborated through a free interpretation[3] of ideas about discursive democracy proposed by Jürgen Habermas, Joshua Cohen and John Dryzek. For each of the two perspectives I deduce a set of processual democratic norms from

---

[3] By free interpretation I mean that I do not engage in a dogmatic interpretation of the criteria that would be derived from a 'true' reading of those authors if they were faced with the same question, but rather use their ideas as inspiration for proposing a set of democratic norms for which I am responsible

the fundamental value assumptions, and I show how these can be operationalised into evaluative democratic criteria vis-à-vis environmental decision processes.

In the second half of the article I attempt to come up with some answers to the democratic problems of environmental agreements mentioned above. By employing the developed democratic criteria in an analysis of the practical experiences with environmental agreements in Austria, Denmark and the Netherlands I will exemplify how this new form of environmental governance actually performs with respect to democratic qualities.

10.2 Fairness of participation

The liberal theory of democracy is not only that of representative democracy. Certain authors within the liberal tradition have sought -on the basis of fundamental liberal values- to develop fine-grained, normative criteria for democratic processes in general.

Although not mainly occupied with the problem of democracy, Rawls' normative theory of *justice as fairness* (Rawls, 1962; 1988) gives quite detailed prescriptions as to what kind of normative principles should provide the basis for democratic institutions. Rawls' main undertaking is the development of a set of normative principles of justice according to which the basic structure of society should be determined. The principles of justice are to guide the assignment and distribution of primary goods in society, such as rights, liberties, powers, opportunities, income, wealth (Rawls, 1988, p.7ff; p.90ff). Since the political constitution is part of the basic structure, and since the corresponding primary goods of political rights, liberties and power, are crucial in determining people's life expectations and opportunities, these matters should also be guided by the principles of justice.

Now, the main idea in Rawls' theory is to derive the principles of justice from what he calls the *original position*. The original position is a hypothetical state of affairs in which free rational men -positioned behind a *veil of ignorance* where they don't know anything about their own talents, life opportunities, and future situation- deliberate on the fundamental terms of future social co-operation (ibid.: pp. 118-195). Rawls argues that in such a state of affairs men will agree to neutral, moral principles they consider *fair* whatever their social position would be. The fair, agreeable principles are those, which provide the best guarantees for the interests of participants, given that they do not know their future social position.

The two general principles of justice resulting from the imagined fair agreement in the original position are as follows: "First: each person is to have an equal right to the most basic liberty compatible with a similar liberty for others. Second: social and economic inequalities are to be arranged so that they are both (a) reasonably expected to be to everyone's advantage, and (b) attached to positions and offices open to all." (Rawls, 1988, p. 60). The second principle defines what sorts of inequalities are permissible; unequal distribution of goods is only acceptable when there is fair equality of opportunity to attain them, and when the inequalities work out to the advantage of even the least favoured.[4]

From this it seems that Rawls' sense of fairness and his corresponding view on justice is informed by two fundamental values -liberty and equality- combined in a subtle way in his two principles of justice. The ideals of liberty and equality, as expressed in the conception of justice as fairness, are also decisive in Rawl's normative ideas about democratic institutions (ibid.: pp. 195-251). In accordance with the first principle, political and civil liberties -the right to vote and to be eligible for public office, freedom of speech and assembly, liberty of conscience, freedom of thought, freedom of the person, right to hold property, etc.- are the fundamental democratic ideals to be embodied in the constitution. However, it is important to note that these liberties are constrained by the ideal of equality; the liberties should always be equal, that is, they should never go beyond the point, where they interfere with the liberties of others. The first principle of justice can be applied to the question of political participation in general:

> The principle of equal liberty, when applied to the political procedure defined by the constitution, I shall refer to as the principle of (equal) participation. It requires that all citizens are to have an equal right to take part in, and to determine the outcome of, the constitutional process that establishes the laws with which we are to comply (ibid.: 221).

The principle of participation prescribes an equal *right* to participate in the *constitutional process*, but it does not indicate that citizens have an equal right to participate in any political decision-making process. In fact, Rawls himself does not touch upon this divergence. However, he makes a distinction between a constitutional level, which is mainly to be guided by the first principle of justice and a legislative level, which is to be guided by the second principle (ibid.: pp. 198-199). Now, it seems reasonable to say that participation in individual decision-making processes is not so much a question of principle liberties, but rather a legislative question on distribution of powers and opportunities in actual political processes. This question is, therefore, to be governed by the second principle of justice, and we will come back to that after dwelling a while on Robert Dahl's democratic ideals.

---

[4] The second principle denotes a break with utilitarianisme where every inequality that improves the net benefit of society is justified -even if it makes the least favoured worse off (Rawls, 1962, pp. 135-136; 1988, pp. 65-80).

In his major work 'Democracy and its Critics', Dahl presents five ideal democratic criteria, which are derived deductively over a series of stages (Dahl, 1989, pp. 83-119; pp. 213-225; see also Poulsen, 1986, pp. 215-225). Also in Dahl's theory, liberty and equality appear at the normative source in the form of value assumptions on personal autonomy and equal worth of citizens. According to Dahl, the justification for democracy springs from an assumption of *equal intrinsic worth,* a moral assumption implying that no individuals, from nature, have a right to force their will on others. Dahl shares this value assumption and transforms it into a political ideal, that the interest of each person should be taken equally into account *-principle of equal consideration of interests* (Dahl, 1989, pp. 85-86). In essence, this principle is a quest for impartiality in political rule, and as such it is fair in the sense of Rawls. However, it is still a weak ideal of political equality: it does not indicate whether inequalities are still permissible within the range of impartiality; and it does not indicate who is to judge the interests of persons (ibid.: pp. 87-88).

Dahls' solution is to add a *presumption of personal autonomy* stating that each person is the best judge of his own interests, unless the contrary can be proved (ibid.: pp. 100-101). Together with the principle of impartiality this leads to a *strong principle of equality* -on the citizens' equal right, and ability, to self-determination:

> "If the good or interests of everyone should be weighed equally, and if each adult person is in general the best judge of his or her good or interest, then every adult member of an association is sufficiently well qualified, taken all around, to participate in making binding collective decisions that affect his or her good or interests, that is to be a *full citizen* of the demos. More specifically, when binding decisions are made, the claims of each citizen as to the laws, rules, policies, etc. to be adopted must be counted as valid and equally valid." (ibid.: p. 105).

Rawls' and Dahl's normative considerations enables me to phrase a more precise question about democratic norms: aiming at the fundamental values of liberty and equality, or rather their balanced expression in the notion of fairness; what participatory qualities would we want from an ideal process of making collectively binding decisions? Or simply: what should be the norms for fair democratic participation?

The first thing to settle is who should participate in ordinary political decision-making processes. One of Dahl's five criteria is that "the demos must include all adult members of the association except transients and persons proved to be mentally defective" (ibid.: p. 129). Furthermore, Rawls' principle of participation requires that every citizen have an equal right to take part in constitutional processes. But as we already indicated above, this principle, and Dahl's criterion of inclusiveness, does not necessarily give rise to the claim that everyone should have the right to participate actively in any ordinary political process. For the sake of efficiency it must be considered fair enough that only some people be included as active participants, as long as these participatory inequalities satisfy Rawls' second principle of justice. I will put the following way: if we are positioned behind the veil of ignorance, we will agree to certain inequalities in participation, provided that those inequalities can be presumed not to worsen the life expectations of those who will come to belong to weak private interests at the time of determination of their social position. The requirement that all declared *stakeholders* be included as active participants is the only criterion which satisfies this condition. Only by allowing, in principle, for the inclusion of all private parties[5], who rightly claim to have a stake in the policy matter, can we be sure that the expectations of the least favoured stakeholders are not worsened - and the exclusion of others (non-stakeholders) causes no problems from that point of view, but improves the efficiency of the political process in question. We then have our first criterion for fair participation:

(1) *any private party with an express, legitimate stake in the policy matter to be decided upon, should have the opportunity to be included as an active participant in the political decision process.*

Before we go on to operationalise the concept of express, legitimate stakeholders vis-à-vis environmental decision processes we will dwell for a while on an additional criterion for fair democratic processes. The question is how to define more closely the included private parties' degree of participation in the various phases of political decision-making processes. Here we can make use of Dahl's criterion for *effective participation*: "Throughout the process of making binding decisions, citizens ought to have an adequate opportunity, and an equal opportunity, for expressing their preferences as to the final outcome. They must have adequate and equal opportunities for placing questions on the agenda and for expressing reasons for endorsing one outcome rather than another" (ibid.: 109). Dahl stresses equal opportunities for adequate participation in all major decision phases, and his criterion therefore satisfies our requirement of fairness. Our second criterion for fair participation can therefore be formulated as follows:

(2) *included private parties must have adequate opportunities, and equal opportunities, to participate in all phases of the political decision process.*

---

[5] We say *private parties* to indicate that the political process is a sort of game, in which it is fair to include not only private citizens, but also private interest groups and firms.

Operationalisation

The operationalisation of our first criterion -inclusion of all private parties with an express, legitimate stake in the policy matter- will be treated at some lenght, since the definition on who should have the opportunity to be included in environmental decision processes is rather fundamental to the other aspects of these processes. First of all, our criterion refers to *private* parties. It is more or less trivial that public parties (environmental bureaucrats, government officials) participate, whereas the liberty concern for fairness stress the importance of citizens and their representatives (interest groups of political parties) having a share in government. In our operationalisation of the first criterion we can therefore ignore the participation of governing elites.

In the designation of the range of stakeholders in environmental decision processes it seems appropriate to distinguish between professionals and amateurs (cf. Fiorino, 1996). *Elected political representatives* should be counted as one type of professional stakeholders - at least when they act in capacity of party spokesmen on environmental matters. Another type of professional private parties are representatives from interest groups. Interest representatives on environmental matters can roughly be subdivided into three major stakeholder groupings and one remaining category: one major grouping consists largely of *business interests* whose economic activities are the cause of pollution- and therefore made the subject environmental policy decisions. For business interests the essential political game is about minimizing economic and other burdens, following from those policy decisions. Another major grouping is dominated by *environmental- and consumer groups* who engage in political struggles for strict environmental standards for the sake of environmental protection. A third grouping is *trade unions* pursuing the twin objectives of material and immaterial well-being of the workers, the first stressing economically sound conditions for the companies and their workers, the second stressing working environment and overall environmental quality. Fourth, it must not be forgotten that sometimes *other interest groups* also have a stake in environmental policy decisions, for instance scientific, religious and charitable foundations as well as organisations representing native populations on the territory (ibid.: p. 257; Hayward, 1995). Finally, we have the amateur stakeholders, that is, *private citizens* who claim to have a fundamental stake in the environmental policy matter in question.

In order to operationalise the first democratic criterion we now have to decide which stakeholders ought to be included as active participants in environmental decision processes. From Rawls we have learned that the fair solution is that which we all can agree to behind a veil of ignorance. Elaborating on his logic, in the original position none of us would know if in the future we would belong to the powerful interests in the environmental field (insider politicians, business interests) or to the less powerful interests (environmental groups, consumer groups, trade unions, private citizens). Under those conditions we can agree to the fact that it is only fair that any of the above mentioned stakeholders should be offered the opportunity to participate actively in the concrete environmental decision process in which they have a stake. However, we can also agree that it is only necessary to include those stakeholders who expressly declare their interest to participate. Assuming that the parties know their objective interests, and that it is relatively cost-free to declare those, it must be considered fair to exclude those parties who do not express any interests in the policy matter. Accordingly, the basic democratic qualities of an environmental process is determined by the extent to which it includes all political representatives, business organisations, trade unions, environmental groups, consumer groups, private citizens, and other groups with an express stake in the policy matter.

However, the suggested operationalisation of the fairness requirement into a practical criterion for inclusion in environmental processes runs into a major problem. It relates to the very nature of the environmental problems, in particular the plea for sustainable development and its moral imperative not to compromise the needs of future generations. When we are behind Rawls' veil of ignorance we do not even know to which generation we belong (Rawls, 1988, p. 137). The logical implication is that future generations are also legitimate stakeholders who somehow ought to be included in environmental processes. However, the problem is that the instrumental devices of liberal democratic theory -namely pressure groups and citizen demands- does not provide us with a solution. Of course some would argue that environmental groups raise the voice of future generations through their struggle for environmental protection. Others would argue that state actors will take care, or at least that neutral experts will raise the question of future needs. But one cannot escape the fact that all these actors are financed, elected or supported by current constituencies, whose needs are of primary concern. I argue that the problem is unsolvable from the point of view of liberal democracy, and this already indicates the need for alternative criteria of democracy.[6]

After this lengthy discussion we only need to say a few words about the operationalisation of our second criteria for fair participation. The second criteria is about the extent of participation and it stated that "included private parties must have adequate opportunities, and equal opportunities, to participate in all phases

---

[6] Authors within the eco-centric perspective on the environment have pointed to another problem of similar character: how to ensure that the interests of other living beings (animals, plants. etc.) are democratically represented in political processes (see e.g. Dryzek, 1995; Matthews, 1995; Barns, 1995; Plumwood, 1995; Eckersley, 1995). Although I reject the eco-centric perspective on epistemological grounds (after all, only human beings are capable of value reflections and political decision-making), I do agree on the point that we need something more than liberal democratic practises in order to safeguard 'animal rights' and conserve valuable biospheres.

of the political decision process". Accordingly, fair participation is dependent upon the fact that private parties have an opportunity to contribute in all phases of the environmental process, and not just, for example, in policy-formulation, only to be circumscribed by an already set agenda for the environmental problem, or only to have the decisions altered during the process of implementation. We can operationalise the criterion by asking "Do the included private parties have adequate and equal opportunities to participate in (a) agenda-setting; (b) policy-formulation; and (c) implementation?" To the extent that the answer is affirmative to all three questions, the environmental process comes close to meet the criterion of equal participation.

A very delicate question is how to value citizen participation vs. the various forms of professional private participation. There is no definite answer to this question.[7]

All we can say is that all the stakeholders specified above ought to be included as active participants: (1) the more of those stakeholders included, and (2) the more equal the extent of their participation, the closer the environmental process comes to fair democratic participation.

10.3 Communicative qualities of participation

The ideal criteria for fair participation does not, however, completely satisfy all our possible wishes for democratic participation. They go a long way in transforming value assumptions of equality and liberty into norms for fair participation. Nevertheless, fairness does not pre-empt the meaning of liberty and equality, and more importantly, it does not describe whether and how values relating to the community aspect of participation should be furthered through the democratic process.

It is, therefore, time to look at the community aspects of democracy. However, it is important to note that -somewhat simplified- there are two different approaches within this view on democracy (cf. Habermas, 1992, pp. 324-367). One is referred to as republican/communitarian. This approach emphasises direct participation of all citizens in deciding the common good, considered as the realisation of community's true ethical values. The other -the discursive approach to democracy- is more realistic, proceduralistic and modern. For one, it accepts

---

[7] It is, however, a fact that professional interest representation very seldom is based on internal democratic structures. This goes especially for private interests organizations, including many environmental organizations (Bichsel, 1996: p. 238), where it is most often the case that a small leader group formulates policies and conduct negotatiations. This suggests that higher weight should be given to the inclusion of amateurs, in that participation by ordinary citizens in intself is a genuine democratic feature. But it is impossible to say anything about how much more weight.

the existence of a pluralist society in which people have conflicting interests. Second, it attempts to develop a theory of deliberative democratic processes oriented towards communicative rationality -to be understood as the reasoned outcome of free public debate- but not towards the realisation of substantive ethical values embedded in the existing community.

> [Dem Republikanismus gegenüber] beharrt eine diskurstheoretische Deutung darauf, dass die demokratische Willensbildung ihre legitimierende Kraft nicht vorgängig aus der Konvergenz eingelebter sittlicher Überzeugungen zieht, sondern aus Kommunikations-voraussetzungen und Verfahren, die im Process der Beratung die besseren Argumente zum zuge kommen lassen (Habermas, 1992, p. 339).

With a view to formulate a further set of criteria for democratic participation I will make use of this discursive approach to democracy -proposed by, among others, Jürgen Habermas, Josua Cohen, and John Dryzek- rather than the communitarian approach. The overly idealistic nature of the communitarian approach simply makes it ill-suited to define the processual norms for modern environmental policy-making.

The lesson to be learnt from the discursive approach to democracy is that it is not sufficient to ask whether participation is fairly liberal and equal from a subjective point of view. On the contrary, the discursive ideal of deliberative democracy emphasises the *intersubjective qualities of participation*. Accordingly, political processes ought to be institutionalised in a way that furthers *communicative virtues* of participation, that is, dominance-free argumentative reasoning between communicative competent participants (Habermas, 1981; Dryzek, 1990, pp. 36-38). A structuring of participation along these lines -and this is the central value assumption of discursive democracy- will help to bring about political decisions, which are *reasonable* from an intersubjective point of view, that is, decisions which can be assumed to further the *common interests* of citizens:

> Dessen pointe besteht nähmlich darin, dass das demokratische Verfahren Diskurse und Verhandlungen mit hilfe von Kommunikationsformen institutionalisiert, die für alle verfahrenskonform erzielten Ergebnisse die Vermutung der Vernünftigkeit begründen sollen. ...Die deliberative Politik gewinnt ihre legitimierende Kraft aus der Diskursiven Struktur einer Meinungs- und Willensbildung, die ihre sozialintegrative Funktion nur dank der erwartung einer vernünftigen *Qualität* ihrer Ergebnisse erfüllen kann (Habermas, 1992, pp. 368-369).

All this suggests that the discursive level of public debate is the critical variable in our search for further norms to define ideal participation in ordinary political decision-making processes. Our question then is: Aiming at free public deliberation proceeding in the spirit of communicative rationality, what participatory qualities would we want from a process of making collectively binding decisions? It is already clear that we want participation to exhibit certain communicative qualities, which enhance the prospects for the unfolding of communicative rationality. We can now go more into details about the form and content of these communicative qualities of participation.

The first thing to consider is the mode of participation. In his attempt to transform the ideal of deliberative democracy into an ideal deliberative procedure, Cohen suggests a number of procedural norms. Regarding the mode of participation to be followed by the parties, the norm is that: "Deliberation is *reasoned* in that the parties to it are required to state their reasons for advancing proposals, supporting them or criticising them. They give reasons with the expectation that those reasons (and not, for example, their power) will settle the fate of their proposal." (Cohen, 1989, p. 22). Thus, decision-making processes are in reach of discursive democratic qualities when they proceed through extensive discussions in which participants reach consensual decisions only on basis of "den zwanglosen zwang des besseren Argument (Habermas, 1992, p. 370). On the contrary, communicative qualities of participation are in deficit, whenever deliberation between the parties is merely a kind of bargaining, i.e. where the exercise of dominance relations and power positions becomes the driving force behind agreements. Accordingly:

(3) *the mode of participation should be reasoned deliberation, that is, political decision-making processes ought to proceed through extensive discussions between the parties in which no force other than that of the better argument is exercised.*

Another thing to consider is the locus of participation. The proponents of deliberative democracy emphasise that the ideal place to develop the political potentials of communicative rationality is an open public sphere. Discussions in the public sphere is supposed to be relatively undistorted by instrumental rationality, since participants freely argue on the basis of civil society life experiences, rather than strategic motives. Dryzek takes the point very far, claiming that: "Hence the proper location for any discursive designs is the public space between individuals and the state." (Dryzek, 1990, p. 40). According to Dryzek, discursive democratic processes should be located beyond the domain of the state in what he calls authentic public spheres. In my view, this rejectionist approach is not very realistic, nor is it particularly desirable from a democratic point of view. Political decision-making processes must be formalised to a certain extent and certain structures of authority need to be institutionalised. Otherwise it is not possible to uphold the rule of law and ensure a binding character of public decisions. Hence, unrestricted public debates to which anyone has access should never *replace* formalised processes for making collectively binding decisions. This would only bring about democratic anarchy.

Instead, I am more in line with Habermas. He recognises the need for expertise and a certain measure of instrumental rationality in political processes just as he reflects on the proper balance between the judicial structure of a democratic state and the legitimacy of its basic decision processes. Thus, he maintains that state-institutionalised decision processes are a proper locus for political participation, but at the same time he emphasises that such processes ought to be coupled to a broader public sphere emanating from civil society. This is because: "...die algemeine Öffentlichkeit... hat den Vorzug eines Medium uneingeschränkter Kommunikation, in dem neue Problemlagen sensitiver wahrgenommen, Selbstverständiggungsdiskurse breiter und expressiver geführt, kollektive Identitäten und Bedürfnisinterpretationen ungezwungener artikuliert werden können als in den Verfahrensregulierten Öffentlichkeiten." (Habermas, 1992, p. 374). The implication of Habermas' argument is that communicative rationality will only unfold when political processes are thoroughly linked to a public sphere emanating from civil society (cf. ibid.: pp. 362, 432, 434, 460, 463). This brings us to our fourth criterion:

(4) *there must be a flow of communication between formal participants and the broader public. All major information must sift to the public sphere and the formal participants must take account of new issues, demands and ideas raised in the public sphere.*

Operationalisation
In relation to operationalisation of the democratic norms emphasising communicative qualities, a first observation is that it will be difficult to evaluate environmental processes according to the third criterion on reasoned deliberation. Even though environmental questions are said to have particular good potentials for reasoned deliberation due to their impact on the common faith of communities (Dryzek, 1987), it is hard to collect evidence to this fact. This is because collecting evidence on reasoned deliberation requires detailed reportings of actual negotiation processes most often taking place behind closed doors. Anyway, I will show that it is possible to set out yardsticks for this criterion, which *can* be applied in actual evaluations, although it requires significant efforts to get behind the scene.

One yardstick is to ask whether the environmental process under examination is characterised by the absence of any use, or threats, of force by parties anxious to have their will. In an environmental context force may be exercised in many ways, for example by business interests exercising financial threats of relocating their activities; withdrawing from investments, etc.; by environmental authorities making threats of legislation, fines or use of physical force; by environmental- and consumer groups threatening to organise boycotts of specific products, etc. The absence of such force is so to say the basic practical criterion for environmental processes proceeding on the basis of reasoned deliberation. Additional yardsticks indicate the further extent to which environmental processes are characterised by reasoned deliberation: one indicator is whether some of the private parties attempt to persuade the others through the use of reasoned argumentation referring to common interests in the environment, the economy, etc. To the extent this reasoned argumentation succeeds in moving the preferences of other actors away from their narrow private interests we have further evidence of the presence of reasoned deliberation. The more possitive participation in environmental decision processes turn out on these yardsticks, the more democratic it is from the point view of discursive democracy.

Our fourth criterion stated that "there must be a flow of communication between formal participants and a broader public sphere. All major information must sift to the public sphere and the formal participants must take account of new issues, demands and ideas raised in the public sphere." This is an extremely important criterion because it helps to differentiate between closed corporatist decision structures, where almost all information is reserved for elites and professionals, and other decision structures, which nevertheless maintain communicative interaction with the broader public. Our democratic requirement is that there must be a flow of communication in two directions: *information* must flow to the broader public, and *opinions* must flow from the broader public and back into the formal decision structures. The first flow is important because it allows citizens to formulate critical judgements on environmental policy measures under negotiation -and the ones already decided- from the viewpoint of civil society experience. This will enable citizens to voice their own opinions on environmental problems to the policy-makers, to the extent channels exists for this purpose. These opinions are vital for the flourishing of communicative rationality within environmental decision processes: not only do the influx of public opinions help to clarify civil values on the environment, they also lead to spontaneous formations of new environmental policy agendas bringing about critical issues, which would not have been considered otherwise.

In the operationalisation it is therefore relevant to address the question of *publicity*. First we must ask whether the broader public has access to information on the three phases of formal environmental decision-making (agenda-setting, policy formulation, and implementation). Second, it should be asked whether formal participants actively engage in serious efforts to distribute information to the public, hence, creating a need for information on technically complicated environmental problems in a language comprehensible to ordinary citizens. Remaining are the questions of a flow of opinions the other way around. In that respect we should at least ask, whether formal participants in the environmental process openly make references to public opinions expressed in the course of the process. It is also relevant to ask, whether specific institutional devices are set up to channel public opinions into the environmental decision process, and whether such channels are relatively open to a broader public.

10.4 Evaluation of the democratic qualities of environmental agreements: the examples of Austria, Denmark and the Netherlands

Environmental agreements are written contracts between one or more public and private parties in which those parties voluntarily commit themselves to take environmental measures with a view to realise certain environmental policy objectives. Environmental agreements can be more or less legally binding, but they always stipulate mutual environmental obligations with social consequences for the allocation of environmental, economic and other resources. Hence, they must be seen as collective policy decisions belonging to a particular mode of environmental governance, which we have chosen to term co-operative management. The most prevalent form of environmental agreements (hereafter EAs) is a contract between the environmental authorities and one or more parties representing the polluting industry, in which the latter commit themselves to take certain actions to prevent or abate pollution, and in return the public authorities make a 'best effort' commitment to withdraw from enacting new environmental legislation on the matter.

Such EAs are not always applied as a clear-cut alternative to traditional regulatory control; often they are merely agreements on how to implement existing legislation, and they are sometimes accompanied by legislation on how to make the agreement workable (Basse, 1997). In any case, the adoption of EAs imply a new form of environmental governance with radical different procedures than traditional regulatory governance. The major difference is the interactive character of the processes surrounding the adoption of EAs, i.e. the fact that certain private parties are actively involved in the entire process of making agreements thereby providing for more intense public-private negotiations throughout the process. However, the the interesting question is how such processes actually perform when held up against clarified democratic norms. Below I will give some tentative answers to this question through an analysis of the democratic qualities of EAs in Austria, Denmark and the Netherlands on the basis of the democratic criteria I developed in the preceding sections. I will only highlight some central findings from our ongoing cross-national research project (see note 1), where we examine the general praxis of EAs in the three countries as well as the praxis of EAs in the fields of packaging waste and energy efficiency in more detail. Since the praxis of EAs is much more extensive and complex in the Netherlands than in the two other countries, it will be given a lengthier treatment.

10.5 Environmental agreements and fair participation

The Netherlands.
Along with Japan, the Netherlands has the most developed praxis of environmental agreements in the world. Until 1995, at least 100 EAs were concluded (Algemene Rekenkamer, 1995), but even more impressive than the sheer number is the wide scope of some of those EAs. I will restrict my evaluations to the mature Dutch EAs (cf. Glasbergen, ch. 7 in this book) entered in 1990s after the invention of the target group approach and the adoption of more systematic codes of conduct for the conclusion of EAs.

The most extensive program to date is the Dutch *target group approach* which was adopted in order to translate the medium-term (1995-2000) and long-term (2010) environmental quality objectives of the first and second National Environmental Policy Plan (often called NEPP and NEPP+) into specific emission reduction targets for entire industrial sectors. The central procedural means of the target group approach is the negotiation and conclusion of agreements between on the one hand all the involved environmental authorities, and on the other hand branch organisations representing the targeted business sectors (see also Glasbergen, ch.7; Peters, 1993; Biekart, 1995; Suurland, 1994; Bastmeijer, 1996). The agreements take the form of integrated target plans specifying obligations in relation to multiple pollution sources such as air emissions, waste water, soil contamination, waste disposal, noise, etc. For homogeneous sectors, the branch organisations negotiate a general implementation plan for the sector as a whole, which then suffice as a basis for granting environmental permits to any company within the sector. For heterogeneous sectors the agreement on an overall implementation plan must also be signed by most individual companies, who in turn must draw up company environmental plans (every four years) forming the basis for individual granting of permits.

Now, to what extent can we speak about fair democratic participation in those target group policy processes? Focusing on the negotiation and conclusion of EAs as the very core of those processes there is a democratic problem of non-inclusion of private parties with an express, legitimate stake in the policy matter: only governmental organisations and business interests are included as active participants in the making of Dutch target group agreements, whereas all other stakeholders (elected politicians, environmental groups, trade unions, private citizens, etc.) are left out.

Keeping in mind, however, that the making of agreements is only one step in the target group process, the problem appears in a different light: the agreements are preceded by negotiation and formulation of overall environmental targets for every sector. In this phase of the process the Dutch parliament is involved and a broad range of stakeholders are heard (excluding only private citizens). The importance of this phase is underlined by the fact that the decided environmental objectives are translated into specific emission reduction targets having the status of non-negotiable targets for the superseding agreements. Thus, the element of unfair participation applies more narrowly to the decisions on how to implement sectoral targets, thereunder decisions about time frames, distribution of environmental obligations within the sectors, practical organisation, monitoring- and reporting obligations, etc. Of course these are crucial decisions which could lead to all sorts of circumventions of the environmental objectives in the absence of broader stakeholder participation. The risk that implementation bias the whole process is particularly high due to the relatively weaker means of enforcement compared to direct regulation (the target agreements are civil law contracts without any specified sanctions). However, the broader range of stakeholders still have the possibility to intervene in the following permit procedure, which in the Netherlands is open to comments and objections from any citizen or interest group. Since the signing of a target group agreement does not automatically guarantee that a pollution permit is granted (cf. Peters, 1993; van den Broek, 1993; van Buuren, 1993) the permit procedure reintroduces a certain fairness at the evaluative phase of the environmental decision process.

Far from all recent EAs in the Netherlands are part of the target group approach. Some of the other recent agreements respect the general environmental objectives set by the government and the parliament; others extend to the negotiation over environmental targets, thereby limiting the influence of the sovereign. In any case, most of these EAs are in approximate compliance with the code of conduct for concluding EAs which was finalised in 1992 in order to invoke a more formalised structure on the content and procedure of EAs (Bastmeijer, 1994; 1996). Regarding procedure, the code of conduct specifies that the parliament must be informed about the government's intention to conclude an agreement and that a copy of the final agreement must be sent to parliament. Moreover it states more vaguely that the government *may* decide to submit a draft agreement to parliament if there is ground for doing so. In 1995, under an evaluation of the existing EAs the Dutch Court of Audit found that in 65 per cent of the cases the parliament was informed before the signing of an agreement, and in about half of those cases it received a draft agreement (Algemene Rekenkamer, 1995). The code of conduct does not, however, require the inclusion of third-party stakeholders. It only states it should be indicated "whether, to what extent and at what point in the preparation of the covenant consultations were held with third parties".

With the exception of the agreement on tropical timber (see Glasbergen, ch.7 this book), none of the recent agreements included environmental groups, consumer groups or other weak stakeholders as signatories. Whether they were included at all in the process seems to depend on whether the public initiative for concluding agreements was in the hands of the Ministry of Housing, Spatial Planning and Environment (VROM) or the Ministry of Economic Affairs (EZ). EZ has often opted for closed decision structures, for example in case of the very important long-term agreements on industrial energy efficiency, where no other private parties than industrial federations, branch organisations, and firms, together with NOVEM[8], are allowed to participate at any phase of the process. VROM, on the contrary, has been more keen to involve third-parties, especially environmental groups, in the process. One example is the comprehensive packaging waste agreement of 1991, where consumer- and environmental groups participated in the preliminary strategic talks. However, even though the participation of these groups were financially supported by VROM, they decided to withdraw from the negotiations, especially because they did not wish to loose their independent position by committing themselves to the outcome of negotiations. Thus, here we cannot speak about an unfair exclusion of legitimate stakeholders, because the latter seized to express a preference for participation. Nevertheless, VROM continued to inform and consult the environmental groups until the end of the process and they were granted an observer status in the follow-up committee responsible for monitoring the implementation of the agreement. This procedural praxis is more or less reiterated in other recent agreements for which VROM is responsible.

It is difficult to reach a clear-cut conclusion with respect to the fairness of Dutch EAs. One cannot escape the fact that they are arranged as bilateral negotiations between the government and only one type of private parties -associations or firms representing business interests- with the consequence that other legitimate stakeholders are excluded from important decisions. We can conclude that the problem of unfair democratic participation is particularly urgent in the Dutch cases where bilateral co-operation extends even to decisions about environmental targets, or where the environmental targets are circumvented

---

[8] NOVEM (Dutch Organization for Energy and Environment) is a private organization funded by EZ and VROM in order to carry out environmental services for the business community. With respect to the long-term agreements on energy-efficiency it carries out secretarial tasks especially in relation to implementation and monitoring, but it also assists the industrial parties in preparation and negotiation.

through lax implementation efforts following from soft obligations in agreements. On the contrary, in the Dutch cases where (a) environmental targets are fixed prior to the bilateral negotiations through the involvement of the parliament and a broader range of stakeholders, and (b) the latter are consulted in the course of the bilateral negotiations, and later involved in follow-up groups, and (c) where the public at large retain the option to object to the related permits, EAs are even more fair than traditional regulation, because they allow for more extensive private participation without compromising equality.

Denmark

The Danish EAs are more informal than the mature Dutch EAs. Despite the fact that the Danish Environmental Protection Act contains a well elaborated section on the procedure and content of EAs it has never been used. Neither the industry, nor the government, have shown any interest in concluding legally binding agreements, and they have, therefore, opted for "gentleman agreements" with no juridically binding content. The lack of formality is reflected by several factors: there is no reference to a code of conduct for entering EAs; the Danish parliament is not systematically involved in target-setting, neither is it informed on the negotiations over EAs; there is no link between EAs and the process of issuing pollution permits. When this is said, it should be noted, however, that most Danish EAs are clear with respect to quantified environmental targets, time frames, distribution of responsibilities, and the terms for withdrawing from agreements.

As in the Netherlands, the negotiation and signing of agreements involves only environmental authorities and business interests, but in the light of the informality of the procedure this becomes much more critical from the point of view of fair democratic participation. In Denmark, third party stakeholders only very seldom have the possibility to influence the targets and conditions of agreements, since these matters are decided outside the parliamentary realm. Sometimes a few representatives from trade unions and environmental- and consumer groups are consulted before the preparation of an agreement. But usually those stakeholders are excluded at such an early phase that they do not even participate in the formulation of targets, nor in the negotiations, nor in the signing, at not even in the follow-up groups.[9]

In all the cases where Danish EAs are adopted instead of new regulations (more than 20 examples), a very unfair pattern of participation is implied, since agenda-setting, policy-formulation, and implementation is the privilege of governmental authorities and business interests. Moreover, third parties are

---

[9] In a few cases, however, representatives from the trade unions have participated in the follow-up group. For example the transport packaging waste agreement and the PVC agreement.

deprived of important legal means to bring industrial parties before the Court due to the non-binding character of agreements.

Austria

Looking at Austrian politics, special attention must be paid to its corporatist organisation of public-private interactions, which apply even to new policy areas such as environmental protection (Talos, 1996). The power center of Austrian politics is still the social partners (Business Chamber, Chamber of Agriculture, Chamber of Labour and the Trade Union Federation) seated in the Joint Commission. Austrian policy decisions are most often defined through consensual compromises between these social partners and the two major political parties (the Socialdemocrats and the conservative People's Party) before reaching the Austrian parliament. The informal, consensual style of policy-making and the corporatist structure of interest organisations (obligatory membership, centralised structure) provide very favourable conditions for the practise of EAs between the government and representatives of private industry. On the other hand, the risk is a form of co-operative management which is merely a continuation of elitist, corporatist practises.

In Austria, the Chamber of Business has been involved in all the existing EAs. In most cases the agreements were negotiated by the Chamber in close association with either the concerned industrial subdivision or a voluntary branch organisation of producers. The parliament did not take part in the decisions, nor did it receive information prior to the signing of agreements. Likewise, environmental groups, consumer groups and other weak stakeholders played no role whatsoever in decisions on Austrian EAs. This has to do with the fact that the Ministry of economic Affairs is the responsible authority for all EAs; organising the talks, the ministry has preferred closed, secret negotiations with only representatives for business interests. In fact third parties and the public have no access to any kind of information on Austrian EAs. The allowance of agreements as highly extra-parliamentary, secret means of decision-making should, however, be seen in the light of Austrian EAs usually being very limited in scope (many are single-issue agreements relating to waste disposal). Nevertheless the conclusion is clear: Austrian EAs represent a continuation of corporatist practises in even more narrow, bilateral decision spheres completely sealed off from other stakeholders and the broader public. This is miles away from fair democratic participation.

Conclusion

In general, the examined experiences with EAs give evidence of a rather unfair pattern of participation and thereby a lack of liberal democratic qualities: whereas the liberty and influence on decision-making is strengthened on behalf of the affected business interests, all other legitimate stakeholders play only a very minor role in all phases of decision-making to which the agreement extends. The situation is worst in Austria. It is not much better in Denmark. In the Netherlands the experience is more mixed: in the low range we have the very unfair processes in which third-parties are excluded from participating in any kind of decisions concerning the agreements, and where the conventional permit procedure is weakened due to exemptions. It is not only the old 'immature' agreements that fall within this lower range, but also some of the more recent agreements such as the very important LTAs on industrial energy-efficiency. In the middle range we have the target group agreements where access to information is better (see next section) and where there is somewhat better possibilities for third-parties to intervene in target-setting and subsequent permit procedures. The upper range consists of some special Dutch EAs which include procedural means for making participation less unfair: broader participation in preparation and goal-setting of EAs through active involvement of the Dutch parliament, and involvement of third parties in prior strategic talks and follow-up groups on monitoring and implementation. Those agreements have become more frequent, but do not yet constitute the majority (Algemene Rekenkamer, 1995).

10.6 Environmental agreements and communicative qualities of participation

Evaluating the communicative qualities of EAs, the first thing to consider is to what extent they promote reasoned deliberation. Earlier, I described the difficulties with collecting evidence on that matter, and it should be said that our present empirical evidence on EAs is not sufficiently detailed to allow precise conclusions for any of the countries. Anyway, I will try to identify some general tendencies. In contrast, it is somewhat easier to collect evidence on the communicative qualities of EAs in relation to the fourth criterion -flow of communication between formal participants and the broader public- since it does not require the same degree of detailed insight into negotiation talks. Thus, relatively reliable conclusions can be drawn from our evidence about: public access to information on EAs; the general level of public debate on EAs, and the institutional devices for channelling communication between participants and the public.

The Netherlands
With respect to reasoned deliberation the first checklist question is whether interactions in decision processes are characterised by an absence of threats of force. In principle, this should be one of the hallmarks of co-operative management qua environmental agreements which, by definition, is supposed to

proceed on the basis of communication and dialogue between equal partners (cf. introductory chapter to this book). In general, it seems that experience with EAs in the Netherlands confirm this expectation: in negotiating EAs, the government generally restrain from using its authoritative powers (such as threats on strict legislation if the private parties do not accept its terms), and the private parties tend to accept that they are only able to put limited pressure on the government. In short, the negotiations are very consensus-oriented. But there is more to reasoned deliberation than the absence of threats of force. The two other checklist questions concern whether the parties make use of arguments referring to generalizable, common interests, and whether this actually results in a solution that reflect common interests, rather than a lowest common denominator. Those questions are more difficult to answer.

My tentative answer will start form the observation that the experience is mixed; there are good and bad agreements, or -to be more concrete- there are agreements that are organised so as to promote learning processes, and agreements that are restricted to the promotion of narrow self-interests of one or more of the parties. In the latter case, the parties only see the agreement as a strategic means to avoid stricter regulation. The most typical example of such 'minimalist agreements' (cf. Lauber and Hofer, 1997) is where business interests are preoccupied with avoiding any kind of demanding environmental obligations, being willing to agree only to a set of voluntary measures which allow firms to keep environmental costs and efforts to an absolute minimum. On the other hand, 'learning-oriented agreements' are those where communication and co-operation efforts between public and private parties permit new advances into areas in which there are no obvious answers or ready-make solutions to the environmental problems (Lauber and Hofer, 1997). Learning-oriented agreements lead to a better understanding of the common interests of the parties and society as a whole (in sustainable development) paving the way for innovative solutions to environmental problems. In the Dutch context there are some examples of learning-oriented agreements evidencing strains of reasoned deliberation in bringing about innovative solutions. Although by no means fully effective, the Dutch target group approach has lead to: increased communication between industrial companies on how to tackle environmental problems jointly; innovation of new organisational constructions for better co-ordination and careful monitoring of environmental progress; enhanced search for integrated solutions coping with multiple pollution sources, hereunder increased use of environmental management systems (see also Glasbergen, ch. 7 this book; Biekart, 1995).

On the flow of communication between formal participants and the broader public we must first consider to what extent the Dutch public has access to information on EAs. The code of conduct from 1992 states that the final agreement must be published in the Government Gazette (*Staatscourant*).

Moreover, it states that the Minister concerned *may* decide to announce a draft in the Government Gazette and invite comments (Bastmeijer, 1994). The decision over the latter is of crucial importance for whether the broader public is able to influence agenda-setting and policy-formulation relating to EAs, or not. Unfortunately, the rule seems to be that drafts are not published, but the government sometimes makes other statements in the media relating to negotiation processes over agreements. With respect to implementation and evaluation of agreements better public access to information is generally provided for: by the Dutch Public Access to Information Act, the rule is that the public has legal access to the final EAs as well as to monitoring reports and evaluations on the progress of EAs.[10]

Moreover, according to the target group agreements, company environmental plans are generally open to the public. This implies that the public now has access to large amounts of data on the environmental loads of individual companies, which they did not have earlier (Biekart, 1995). The relative openness surrounding the implementation phase provides at least a minimum guarantee that members of the broader Dutch public can take part in critical evaluations of EAs and exercise political influence on the next series of environmental decisions (being new agreements, revisions of the old, or other policy instruments). Still lacking, however, are specific institutional devices for channelling public opinions into the negotiations over EAs. Thus, in the Netherlands there is so far not much evidence that public debate had any influence on the eventual content of EAs.

Denmark
The absence of threats of force is equally valid for Danish EAs. The deliberate choice of 'gentleman agreements', rather than EAs guided by the Danish Environmental Protection Act, reflect preferences for an informal, consensual atmosphere. In the EAs concluded so far, we have not found much evidence on Danish environmental authorities exercising pressure on the private parties. This is somewhat surprising given the traditional active stance of the Danish Ministry of Environment, and the fact that some agreement processes dragged out over long periods, delaying planned environmental initiatives. The question is whether the relaxed, co-operative atmosphere surrounding Danish EAs has served to promote reasoned deliberation. There are several grounds to doubt this. One is that some Danish EAs have turned out as 'minimalist agreements', watering down

---

[10] One important exception concern the long-term agreements on industrial energy efficiency. Here, the private organisation NOVEM (see note 8) has the final responsibility for all monitoring and evaluation, which means that the documents remain private property. The consequence is that access cannot be gained by way of the Public Access to Information Act. However, the Court held that this construction is contrary to the meaning of the law.

environmental standards without coming up with new innovative solutions. This goes especially for the handful of agreements on agricultural pollution for which the Ministry of Agriculture is responsible, and some of the wide ranging agreements on polluting substances from industrial processes (PVC; VOC) (see also Holm and Klemmensen, 1993; Lorenzen et al., 1994).

Another reason to doubt the flourishing of reasoned deliberation is the fact that there is virtually no flow of communication to and from the broader public. Danish EAs are not published. However, a recent (unofficial) change in administrative praxis has made it possible for citizens to get access to final agreement texts on request. Nevertheless, there is almost no public debate on EAs in Denmark and in general the public does not know about their existence. Moreover, there is no praxis of letting the public have access to monitoring reports drawn up by the industrial parties or verification reports made by the public authorities.[11]

This means that - besides a few governmental notes on the progress of EAs (see for example Ministry of Environment, 1994) - the public has no information about the implementation of Danish EAs. Thus, it is clearly that the deliberation over Danish EAs takes place in a vacuum, far away from civil society viewpoints.

Austria

Like in Denmark, the Austrian EAs are negotiated in an informal, non-coercive atmosphere. However, it is even more difficult to detect elements of reasoned deliberation in the Austrian EAs, since the great majority come close to being 'minimalist agreements'. The many agreements in the field of waste recycling (bottles, batteries, construction material, plastics, tires, car wrecks, asphalt, etc.) provide good examples (Lauber and Hofer, 1997): the agreements, which came to replace planned decrees on recycling, does not have any binding effect. Moreover, they completely skipped the original idea about deposit/refund systems, because business interests found it too complicated and bureacratic. Rather than coming up with innovative solutions in the alternative, the agreements created a market for waste and organisations having a vested interest in the constant flow of waste.

Even more than in the Danish cases, reasoned deliberation over Austrian EAs is impeded by closure towards the broader public. The Austrian public does not even have access to the final agreements; they are kept as secret documents by the Ministry of Economic Affairs. If progress or implementation reports exists at all, they are also treated as confidential information. In fact, the Austrian public is

---

[11] It might be possible for citizens to have access to this information -if they exist in document form (!)- by reference to the very liberal Public Access to Information Act. However, the authorities may claim that the requested information is confidential.

largely ignorant about the existence of agreements. Hence, it is no wonder that there is no public debate about Austrian EAs. There is not much more to say about the communicative qualities of Austrian EAs. The picture is so clearly negative that it hardly warrants further qualification.

Conclusion

As shown in earlier sections, a main point to be derived from Habermas is that the communicative qualities of EAs with respect to reasoned deliberation is critically dependent on input from the broader public. In the Netherlands, there is a basic interaction between those two aspects due to the relative openness of the Dutch EAs. Although a much better flow of communication ought to be attainable, especially in relation to the negotiation processes, the situation is certainly better than in Denmark and far better than in Austria. Nevertheless, the general conclusion is that EAs do not perform very well in relation to the criteria I formulated as a basis for judging the communicative, democratic qualities of this new way of environmental governance.

10.7 Concluding perspectives

Through clarification and elaboration of two sets of fundamental democratic norms following from deep-rooted occidental values of political association, I have tried to answer the questions posed in the beginning about the democratic qualities of co-operative management qua environmental agreements.

One conclusion is that -in general- the practice of EAs do not live up to the requirements of fair democratic participation, emphasising values of liberty and equality. Various legitimate stakeholders -environmental groups, consumer groups, trade unions, and private citizens with a vested, express interest- are systematically excluded from virtually all phases of decision-making. (A notable exception is the scattered Dutch experiences with more inclusive agreements). Some might argue that this is really not a problem, since liberal democracy is anyway an archaic model, unsuited to modern environmental governance. But one need only think of the immense importance ascribed to the values of liberty and equality by ordinary citizens. Several analyses have shown that people ascribe more importance to the fact that political solutions are arrived at through fair processes, than to the justice, or qualities, of the outcome as such (see e.g. Bingham, 1986; Kluegel et al, 1995). Moreover, the ecological modernisation discourse and its applications are marked by a presumption in favour of public participation (Weale, 1992, pp. 167-180). Thus, what is at stake in unfair processes of environmental agreements is the grander issue of the legitimacy of environmental governance.

A second general conclusion is that EAs in practise fall short of the communicative virtues claimed by some of its proponents. Even though co-operative management have potentials in terms of communicative qualities of democracy -due to the consensual, non-coercive setting of agreements- they are usually not realised, because deliberation takes place only in closed circles, where there is too little constraint on the pursuit of narrow self-interests. However, the positive experiences (e.g. the Dutch target group approach) show that co-operative management can be arranged more in the way of learning-oriented processes, if there is more openness towards the public in all phases of decision-making.

The study has also generated a number of more specific conclusions on the democratic qualities of environmental agreements in the three countries under investigation. The relatively better democratic qualities of the Dutch praxis of environmental agreements, compared to the two other countries, are particularly due to the more formalised, systematic approach chosen in the Netherlands. Although the Dutch praxis still has not come up with adequate solutions to a proper inclusion of third-parties in all phases of decision-making, and still leaves something to wish by way of communication with the broader public it is clearly more mature than the Austrian and the Danish praxis. The undemocratic nature of Danish EAs appear somewhat surprising in view of the traditional strong emphasis on equality and democracy in Danish political culture. However, processual concerns have lead to increasing Danish scepticism with regard to EAs. For example, the Ministry of Environment and Energy has not initiated many new EAs under the Socialdemocratic coalition governments, which came to power in 1993. In fact, today the initiation of new EAs has almost come to a temporary stop.

Finally, let me say a few words about possible avenues for improving the democratic qualities of EAs: First of all, I see no insurmountable barrier to the inclusion of environmental- and consumer organisations as parties to EAs. This is common practise with respect to many EAs concluded in the US and Canada (Lewis, 1996; Webb and Morrison, 1996). While one can argue that it might disturb the consensual atmosphere of the negotiations, the counter argument is that in the last decade - at least in the North West European countries - business and environmental groups have cultivated better relations and learned to negotiate with each other in a disciplined manner. Of course, there is the risk that environmental groups refuse to sign EAs out of ideological considerations or in order to preserve freedom of manoeuvre. However, inviting environmental groups to take part in the negotiations in all cases where the subject matter of EAs is of more general public concern would greatly enhance legitimacy, refuting most claims about unfairness. If, in addition, the parliament is given a final say in the setting of sectoral environmental targets to be implemented by way of EAs, and if it is systematically involved through subsequent ratification of all EAs, then EAs would be as democratically fair as one could reasonably wish. While these procedural guarantees would somehow reduce the flexibility of EAs, they would still be more flexible than regulations since the voluntary character and the specific framing in overall targets and wide time-limits would not be challenged.

When it is deemed unfeasible to include environmental groups directly in the negotiations, one can still ensure a basic degree of fairness by letting them (and other interested third-parties) participate in prior strategic talks setting the agenda for subsequent negotiations. Moreover, for democratic reasons it is highly recommendable that one or more third-parties are given the opportunity to participate in follow-up groups / steering groups responsible for implementation and monitoring of the agreement. In many cases this will require governmental funding of the staff and activities employed by third-parties for that purpose. With respect to improving the communicative qualities of EAs, I think that a great deal can be accomplished by way of the more inclusive participation sketched above, because argumentation will be focusing more on common interests in the search for consensus. Moreover, not only for the sake of communicative democratic virtues, but also in order to arrange EAs more in the way of learning processes, there should be more openness around the whole process: first of all it would be a good idea to publish a declaration of intent and invite for public comments before the parties start real negotiations. Second, it is very important that there is full openness around the implementation and evaluation process. In particular I would recommend that the public has access to all monitoring reports, including those made by individual companies. This would enable more public debate and provide for a critical feed-back on individual agreements. The counter argument that it is necessary to keep monitoring reports confidential due to sensible company information do not seem justified when looking at the actual content of the reports.

Part III

Prospects

11. POWER, PARTICIPATION AND PARTNERSHIP

The limits of co-operative environmental management

ANDREW BLOWERS

11.1 Business as usual?

In order to assess the limits of co-operative strategies for environmental management it is necessary to adopt a critical perspective on the strengths and weaknesses of the approach and to offer a more general overview of the process as a whole. Earlier chapters have shown that considerable progress is being made in specific circumstances though often outcomes are uncertain. In this chapter I intend to address in more general terms the question of how far co-operative strategies offer a solution to environmental problems.

Co-operative environmental strategies have emerged at a time when there has been a changing emphasis in the ecological dimension of public policy. This has been in response to a growing awareness, especially in the West, of the potentially catastrophic global economic and social consequences of human impacts on the natural resource base. Long standing problems of soil deterioration and water and air pollution affecting large areas within countries have been joined by transboundary problems which threaten the survival of ecosystems, sometimes on a world scale. Although the possibility of limits to continuing growth has long been recognised, it is only in the past few years that the urgency of the problems of resource depletion and pollution have begun to pervade the policy making process at national and international levels. It has become increasingly evident that contemporary rates of resource extraction threaten, not merely to increase the costs of scarce resources, but, through the loss of biodiversity, to strip the world of vital and irreplaceable assets. Atmospheric pollution is no longer simply a threat to public health in areas of industrial concentration but, through global warming, ultimately endangers the very survival of life on the earth in the longer term. The dangers of widespread radioactivity resulting from a major accident at a nuclear plant cannot be discounted given the number of accidents that have occurred so far; and the prospect of a nuclear war escalating into a global exchange cannot be ruled out especially as the potential sources of such conflict multiply as a result of proliferation. The prospect of environmental catastrophe, sooner or later, cannot be gainsaid.

P. Glasbergen (ed.), Co-operative Environmental Governance, 229–249.

The political response, at least in terms of policy formulation, has, in certain respects, been quite impressive. Most nation states have issued reports on the state of their environment and many have produced environmental policy statements intended to be implemented through the various vertical policy sectors. At the international level, of course, the Rio Earth Summit in 1992 marked the apotheosis of environmental concern firmly establishing the notion of 'sustainable development' as the goal of international policy. Although the prominence given to the environment at that time has since abated (for instance, there was much less media attention devoted to the Rio plus 5 conference in New York in 1997) the Rio process has continued through the elaboration of conventions on biodiversity, global warming and desertification and in the development of Agenda 21 through the UN Commission for Sustainable Development. Moreover, the process itself by involving business, NGOs, local government and other interests has established the idea of partnership in a common enterprise and the idea that solutions can be sought and achieved through negotiation. A similar notion has been embraced by governments which have appealed to business, local government and environmental interests to co-operate in the development of strategies for sustainable development. Throughout the advanced liberal capitalist democracies the concept of negotiated solutions to deal with the environmental question have taken root and flourished. Earlier chapters have underlined the prevailing belief that a consensus can be developed through participation in private/public partnership which, at one and the same time, can secure a *sustainable* environment without impeding economic development.

The notion of co-operative environmental management, especially between the public and private sectors, has been much debated. Indeed, environmental agreements of the type discussed in this volume can be seen as part of a broader approach that has come (by some academic commentators) to be called *ecological modernisation* (EM). Some would claim that EM has acquired the status of a theory in that it introduces ecological criteria into the debate about modernity. But, EM appears to be more a description of the ways in which business and governments have adapted to the need to incorporate ecological issues into the production and policy processes than an explanation of the process of environmental policy development. It serves an ideological purpose, too, in so far as it supports the prevailing structures of power based around the liberal market economy. It is tendentious in its central belief that environmental conservation and economic development are not merely compatible but mutually supportive. As one of its protagonists argues, "Instead of seeing environmental protection as a burden upon the economy the ecological modernist sees it as a potential source for future growth" (Weale, 1992, p. 76). Under ecological modernisation, the goal of sustainable development has become *de rigueur*, the anticipated outcome of policies and processes that take the environment fully into account. Indeed, sustainable development is now almost taken as given so that it has become tedious to pronounce it let alone analyse its meaning.

For the purpose of the analysis which follows, EM is taken as the prevailing mode of environmental decision making in modern Western capitalist economies. While it incorporates elements of all the models of governance outlined in chapter 1, its emphasis on collaborative strategies between business and government in a market economy provides the context for examining the limitations of co-operative strategies which have been the theme of this book. The argument put forward here is that negotiated solutions for environmental policy offer, at best, only a partial solution to the problem of securing sustainable development. This is because such solutions are (implicitly) predicated on the various assumptions of what may be called 'ecological modernisation'. The argument proceeds as follows. In the following section (2) the idea of EM (and of co-operative environmental management as an integral component) is set out in more detail to establish its position as a dominant form of environmental ideology and its partial nature is outlined. In section 3 the UK is used to exemplify the way in which ecological modernisation works as a negotiated form of solution in a national context and the limitations of the process. Section 4 takes the nuclear industry as a case study of an industry which cannot satisfy the credentials of EM. This critique leads to the obvious question - if EM can only partially succeed, how can we achieve sustainable development? A definitive answer is, of course, at this stage in the debate impossible, but certain pointers are put forward in section 5 from which, it is hoped, more creative ideas will fructify.

11.2 Ecological Modernisation - the Ideology of Sustainable Development

A Consensual Relationship
Ecological modernisation, as a theoretical idea, has its origins in the 1980s, beginning in Germany then spreading to the Netherlands (Hajer, 1996; Mol, 1995, 1996), the UK (Weale, 1992) and subsequently achieving widespread currency in the advanced western countries (Christoff, 1996; Dryzeck, 1995; Cohen, 1997; Blowers, 1997a). There are three underlying premises of EM.

The first is *technological*. Perhaps the key feature of EM is that ecological criteria are imported into the whole sequence of processes from resource development through production to consumption and recovery. Thus, life cycle analysis, recycling, pollution prevention, waste minimisation are applied to production and the panoply of environmentally friendly consumption concepts are deployed (recycled packaging, environmentally friendly resources, sustainable timber etc). Science and technology are perceived as benign influences providing innovation and ingenuity in order to minimise environmental damage. Thus, replenishment, rehabilitation, substitution and abatement are universally encouraged. Several examples are discussed elsewhere in this book.

The second premise is *political*. EM is predicated on the existence of the liberal market economy. The market is seen as the mechanism for flexibility, innovation, responsiveness, efficiency and adaptability. In principle, free trade leading to comparative advantage is upheld as the means to ensure resources are used "as efficiently as possible and this helps to save natural resources" (Faber, 1996, p. 83). However, it is recognised that unfettered competition and free trade can often lead to disastrous environmental consequences, hence the free market must be coupled with the 'enabling state'. The state (and in cases of transboundary problems, the international community) must set the rules and regulations to ensure the market operates fairly and that certain standards are met, certain activities (eg dumping) are prevented and that surveillance is achieved through monitoring to stop free loading or illegal actions. Wherever possible policy implementation through market mechanisms is to be preferred to control and sanctions.

Thirdly, there is the premise of *participation*. It is characteristic of EM that the development of environmental policy is seen in terms of a partnership especially between business and government. There is a mutual dependence between business and industry (Lindblom, 1977) with governments needing the wealth generated by industry in order to maintain economic performance and industry requiring government to provide a level playing field and the physical and social infrastructure necessary to ensure a skilled and healthy workforce, transportation, housing and so on.

While the government/industry nexus is pre-eminent, other actors may also be participants in environmental policy making. This is increasingly true of environmental groups, most notably those environmental NGOs which adopt safe, non-confrontational and supportive platforms. To an extent, prominent but critical NGOs such as Greenpeace and Friends of the Earth are also increasingly drawn in to the policy making process as consultees, on committees or as observers. There remain some elements of the environmental social movement which refuse to be co-opted and consequently are marginalised in the decision making process (Sklair, 1994).

Taken together the premises of ecological modernisation provide a description of the way in which the modern western capitalist state approaches ecological issues. It is a green-tinted liberal democracy which celebrates an ideologically conservative view of economic development. The introduction of ecological concerns has promoted certain changes which are contingent on the relationship between the economy and the environment. As Weale observes, "The challenge of ecological modernisation extends..beyond the economic point that a sound environment is a necessary condition for long-term prosperity and it comes to embrace changes in the relationship between the state, its citizens and private corporations, as well as the relationship between states" (pp. 31-32). EM focuses on tackling the source of environmental problems, not merely in dealing with the impacts. It also suggests the idea of partnership between government and industry through which solutions to environmental problems can be negotiated. It embraces also a notion of participation which incorporates potentially hostile elements such as environmental groups into the decision making arena. EM is pragmatic, consensual and, in a sense, apolitical for it engages both the New Right which ushered in the era of the free market ideology and the modernising wing of the Left (distinguished as 'New' Labour in the UK) which espouses the notion of the 'stakeholder' society.

A partial analysis

While EM focuses on technological and economic changes to respond to environmental change, it essentially visualises modernisation as a continuing, evolutionary process. As one critic puts it: "All that is needed...is to fast forward from the polluting industrial society of the past to the new super-industrialised era of the future" (Hannigan, 1994, p. 184). EM is a reformist programme, emphasising institutional adaptation. "It assumes that the existing political institutions can at least give birth to new supranational forms of management." (Hajer, 1996, p. 253). But it is, at best, a slanted and partial description of the nature and prospects for the modern industrial state.

This partly arises from the origins of EM in western (European) democracies. It is not surprising that EM has been most enthusiastically adopted in western Europe, a continent that is heavily industrialised and whose landscapes have long been settled. Here the priority is the reduction of pollution, a problem which EM readily addresses. Elsewhere, the protection of wilderness is of much greater importance. There is very little extensive wilderness left in western Europe. In Britain, only remote parts of Scotland such as the Flow Country, an area of pristine peat bog in Caithness, can claim to be true wilderness. By contrast, in Australia the predominant environmental issues are the protection of native forests, of major environmental assets such as the Great Barrier Reef and of remote areas such as south west Tasmania hitherto untouched by modern development. These are areas which can only be properly conserved if they are left alone. In other parts of the world the protection of habitats is of key concern and the forests of south east Asia and the Amazon have been the focus of international attention.

It is difficult, if not impossible, to apply EM to developing countries and more especially those with predominantly rural, subsistence economies, where poverty is endemic and which are often lacking the institutional capacity to deal with environmental problems. Here, the main environmental problems are those arising from the need for conservation, not modernisation. But, modernisation in the West may contribute to the growing environmental problems of the South. It is the demands for timber and other forest resources that has, in part, caused the deforestation that now affects large areas in south east Asia. Demand for tropical products and the introduction of modern agricultural practices have, in some areas, led to deterioration of land quality. Large-scale projects such as dams have often destabilised ecosystems and displaced populations. Hazardous industries may be developed in 'pollution havens' and hazardous wastes dumped in poorer countries.

But, even if modernisation is the cause, it is hard to see how ecological modernisation (as currently defined) is the solution. It is widely recognised that these problems are complex and require sensitive, often localised solutions which combine a range of economic, social as well as environmental measures. Sustainable development in the South is more likely to be achieved by the alleviation of poverty and the encouragement of indigenous farming systems than it is by the introduction of modern agricultural systems and practices (Sage, 1996).

Even in the West EM does not embrace the whole range of industrial activities. As one of its arch exponents acknowledges, "the main frame of reference of ecological modernisation includes 'normal' environmental problems such as water pollution, chemical waste and acidification." (Mol, 1995, pp. 395-396). It does not deal with problems of wilderness and cannot deal with those industries which create high-consequence risks such as global warming or nuclear proliferation. The simple point is that such industries cannot be ecologically modernised to the point where risk is eliminated. The reality is that EM, by reducing the probabilities of a risk occurring, encourages the persistence and possible expansion of a technology that is inherently dangerous, a point I shall examine in more detail later.

EM also neglects the social context of environmental change. There is an implicit assumption of consensus and even of social cohesion; it is as if the environmental challenge can be solved simply by technological and economic means. Support for EM will be engendered through a process of social and institutional learning whereby environmental protection is an accepted social norm (List, 1996). Of course, EM does not deny that society is plural with varied and different interests. Indeed, the whole dynamic of capitalist markets depends on the propensity for consumers to exercise their different choices in the market place. But, interests in society are not merely different, they are sometimes divergent. There are considerable inequalities in wealth and power and these have been exacerbated with the triumph of the market - they are its natural condition and outcome. A notable feature of western democracies has been the growth of an underclass experiencing insecurity of employment, a withdrawal of welfare provisions, sometimes suffering poverty and alienated from the political process. Far from being cohesive, bound together by shared norms, integrated by social institutions of family, religion, neighbourhood or class, and supported by a state welfare system, society has become fragmented and individualised. In a system of structural inequality the powerful are unwilling to surrender their economic gains while the disadvantaged are bound to claim a greater share of wealth. Thus, economic growth will remain the priority with environmental protection a less pressing commitment. This may cause some resistance to EM where it may impose short-run costs to producers or consumers which they are unwilling to bear.

Social inequality exposes the partial nature of the notions of consensus, partnership and negotiation which are functional to ecological modernisation. It may be argued that conflict is the more natural condition. There is much evidence for this in western societies (and certainly elsewhere, as events in Eastern Europe have shown). The growth of social movements engaged in mobilising support for particular issues and adopting a variety of methods (collaboration, confrontation consciousness raising) is a phenomenon that demonstrates the vitality of forces operating within civil society (Pakulski, 1991; Yearley, 1996; Thomas, 1996). They provide a challenge to the routine and institutionalised norms in society. They present 'recurrent patterns of collective activities which are partially institutionalised, value oriented and anti-systemic in their form and symbolism' (Pakulski, 1996, p. xiv). Environmental movements have been prominent articulating alternative approaches to sustainable development, engaging in campaigns and protests and deploying a variety of tactics in the relationships with government and industry. As was indicated earlier, environmental movements may encourage and become incorporated into the negotiative process of EM; but some remain outside, critical of the outcomes and urging alternative strategies.

Nor is it always the case that government and industry have the same objectives in seeking negotiated solutions. The idealised notion of a free market and an enabling state engaged in partnership and negotiation masks points of friction. The state's enabling role in setting standards and applying regulations with potential sanctions for non-compliance contradicts the idea of self-assessment and self-regulation. Although business and government will negotiate to resolve differences, compromises have to be made which sacrifice environmental protection in favour of economic exigencies. There are numerous illustrations of polluting or dangerous industries where the economic costs of modernisation are too high and the social costs of closure too great.

More importantly, business is constantly pressing government to reduce its tax burden, to provide infrastructure, and to carry the burden of welfare and education for its workforce. Such contradictory pressures can lead to the kinds of fiscal and legitimation crisis predicted by O'Connor (1973) and Habermas (1976) and, in such circumstances, it would not be surprising if environmental regulation was among the first burdens to be removed from industry. Moreover, while governments are likely to give greater consideration to the national interest, businesses are concerned with markets and shareholders and not necessarily confined to one country.

While EM is undoubtedly an important, indeed a vital element in dealing with the environmental problem, it has evident limitations. It can act as as a potential barrier to more fundamental changes. Shifts towards greater ecological concern may be as much tactical as strategic, necessary to reflect scientific and environmental concerns while, at the same time, securing the continued legitimation of prevailing patterns of power. This interpretation can now be elaborated with the example of the United Kingdom.

11.3 The United Kingdom - exemplary or deviant?

A suitable case for ecological modernisation

In many respects the UK appears to exemplify ecological modernisation rather well. After all, the country was the first major western state to put into practice the full blown modern version of the capitalist market economy. The Thatcherite revolution, beginning in 1979 and developing thereafter, is rightly hailed (or reviled depending on political preference) as a major turning point. The basic principles of the New Right - free market, strong state and selective welfare - have been put into effect by a continuing programme of deregulation, privatisation, centralisation, and welfare reforms which have transformed the relationships between business and labour, business and government, and individuals and government (Gamble, 1988; King, 1987). In the process practically all the nationalised industries (steel, water, electricity, gas, airline, and railways) have been privatised; trade union power has been severely reduced; local government has been emasculated; market principles have been introduced into aspects of the health service; welfare provisions have been reformed and, in some areas, withdrawn; business influence in the management of schools and universities has been increased. Everywhere market testing, market signals and market mechanisms are applied. The benefits of all this - in profits, freedom of choice - must be measured against the costs - unemployment, growth of poverty and insecurity. But - the point to be made here - is that the shift from the post-war welfare state settlement to the contemporary model of the liberal market democracy began earlier and has probably proceeded farther in the UK than anywhere else in Western Europe.

The UK ostensibly offers propitious conditions for EM to take root. There are several aspects to this. First, is the *emphasis on the market*, the acclamation of business as the engine of growth and prosperity. The deregulation and the dismantling of restrictive practices throughout the economy introduces a climate of co-operation between government and industry. Of course, relationships have always been close and business has enjoyed privileged access to government. This emphasis is reflected in environmental policy making. The environmental white paper, *This Common Inheritance,* published in 1990, set out the philosophical principles. The belief that 'new investment and environmental improvement often go hand in hand' (HMSO, 1990, p. 7) is redolent of ecological modernisation.

A second aspect is *the idea of partnership.* Here it is not merely partnership between business and government, important thought that is. The notion has been considerably broadened to draw in all the so-called 'stakeholders' in the process. The idea of community based partnerships has become a key feature of the Local Agenda 21 process and involves local government, voluntary organisations and community based groups among others co-operating on a variety of programmes and projects. Apart from attempting to stimulate social and economic regeneration or environmental improvement, such partnerships are intended to increase participation and empowerment. Partnership also has the function of helping to legitimate policy.

Thirdly, there is the *pragmatic approach.* The UK has always had a strongly pragmatic attitude to regulation. Administrative discretion has been the hallmark. In environmental protection this has enabled negotiation to take place between the regulator and individual industries whereby mutually acceptable settlements are made. This procedure offers maximum flexibility within overall environmental constraints and results in varying standards according to local or industrial circumstances. The traditional Best Practicable Means (BPM) approach enabled economic factors to be taken into account when applying pollution controls. By contrast, continental countries have tended to rely on a more regulated, top down and standards driven form of environmental policy which provides for consistency of outcome. Of course, the contrast between the two systems is more apparent than real and, in any case, has been much reduced with the advent of common policies in the EU. Nevertheless, the difference of regulatory culture still persists suggesting that the UK offers a comparatively permissive climate for EM.

Fourthly, the UK has a *highly centralised political structure.* Indeed, it is politically one of the most centralised unitary states in the western world. Local government, lacking any independent constitutional basis, has been all but eliminated as a political counter force over the past two decades. Political power has become highly concentrated in the executive. The bureaucracy is, on the whole, subservient, disinterested and efficient. The political environment is therefore eminently capable of ensuring that negotiated solutions are carried into effect. It is a relatively simple and, in terms of implementation, effective system.

Paradoxically, although political power is concentrated, it is, at the same time, also dispersed. This reflects a fifth characteristic, the *decentralisation of decision making*. Large areas of public administration have come under the aegis of quangos (quasi-autonomous non governmental organisations) staffed by appointees drawn from various groups including academics, local government, the voluntary sector and lay people. But appointments from business have been a key feature since business is now expected to play an active role in social as well as economic policy. Business people have been appointed to a raft of quangos ranging from health trusts, through school governing boards, to police committees, and development corporations. They are also prominent on boards of privatised utilities, policy advisory committees and on environmental bodies and agencies. Thus, decentralisation of power, if anything, has reinforced the influence of business on policy making and implementation, again a feature which should assist the process of ecological modernisation.

Finally, a sixth point, the UK has a *highly developed civil society*. In some respects, this compensates for its relatively undemocratic political system. The withdrawal or privatisation of many state functions there has resulted in what Rhodes (1994) has described as the 'hollowing out of the state'. Political parties have lost ground as sources of influence for a combination of reasons. Among these are the diminishing influence of local government and Parliament; the convergence of parties over many policy areas as they seek out the 'middle ground'; and the consequent loss of allegiance as party differences are perceived to be more rhetorical than real. The recent (1997) general election was a contest of management rather than ideology. This combination of state withdrawal and loss of political influence has left (or created) a void or political space which is increasingly occupied by interests and movements seeking to influence policy making through consciousness raising, persuasion, lobbying, mobilisation of opinion or direct action. This is what Beck (1992) describes as the 'zone of sub-politics' which, increasingly, has become the arena for developing support for particular issues and policies which are transmitted to the executive through various means of co-operation and consultation. In terms of EM, potentially dissenting viewpoints can be converted through a process of co-option which both broadens the scope of participation while also legitimating policy outcomes.

A sceptical approach

For these various reasons the UK appears to offer optimal conditions for the development and delivery of practices and policies of ecological modernisation. At the rhetorical level of policy making, both in industry and government, EM is deeply entrenched. Yet, in terms of practice it is clear that the UK lags behind countries such as Germany, Japan and the Netherlands. Not only does it lag behind there are good reasons for doubting whether ecological modernisation has had the transforming effect that its supporters claim. A key feature of EM is that it regards environmental protection and economic development as mutually compatible; indeed, it often goes further in asserting that ecological concerns are a necessary condition for economic development. There is some truth in this. It is evident that those countries which have invested heavily in pollution abating and resource conservation technologies have reaped some economic benefits in efficient technology and the sale of pollution control equipment. They are also able to benefit as environmental regulation tightens thus eliminating industries unable to meet the standards. In this sense environmental regulation is a factor influencing competition. In the long run comparative advantage may be achieved by those countries and industries which have taken an early lead in investing in EM. As time goes on more and more economies will recognise the benefits and the necessity for adopting EM strategies. It is not self-evident that EM is an inevitable outcome. EM, at least in the short run, represents an added cost which results in diminished profits or higher prices and less competitiveness. Business will evade the imposition of such constraints if it can, just as it seeks to lower costs through cheaper labour. In order to avoid the extra burden of EM, business may seek out locations elsewhere, often in the South, where investment is welcomed and where regulations are less onerous or lax (Christoff, 1996). It is also true that countries may procrastinate over the imposition of controls in the full knowledge that pollution burdens may be exported. For example, the UK for long refused to accept responsibility for acidification in the Nordic countries as a result of airborne pollution being carried eastwards by the prevailing winds. In the case of global warming the major contributors to greenhouse gases continue to enjoy economic benefits while the costs are dispersed around the globe. Although the long run damage may be terminal, individual countries continue to enjoy a short run benefit - a clear case of the Tragedy of the Commons.

At the political level the EM model with its emphasis on co-operation, negotiation, partnership and consensus stands in sharp contrast to a view of British politics as unequal and conflictual. This goes beyond the adversarial nature of political discourse in the UK (a product of a tendency to strong one party government by contrast with many continental countries where coalition is the norm or where federal systems provide checks on central authority) to a more profound condition of exclusive and elitist decision making. By its nature British government is secretive, lacking a freedom of information act (though one is now in prospect). Although there has been greater openness, some of it pushed through reforms such as the latest by Nolan, there is an almost pathological tendency for politicians and bureaucrats to maintain secrecy. This political culture is distinctive and results in much routine decision making remaining hidden, often in the 'public interest'. As a result, politicians are much less accessible than in many other countries, bureaucrats keep a low public profile and revelation is secured through an investigative media and a culture of leaks.

In certain respects the exercise of power in the UK conforms to a neo-elitist interpretation of power (Bachrach and Baratz, 1970; Crenson, 1971; Blowers, 1984). Among its most obvious characteristics are that power is concentrated among political, bureaucratic and economic elites, access to the power structure is restricted and decision making is hidden. Among the groups enjoying privileged access are business elites. Although other interests such as environmental groups gain admittance from time to time, are occasionally appointed to quangos or to advisory bodies and are routinely consulted, their access is intermittent and specific rather than privileged and routine.

The absence (on an equal footing) from the power structure of interests promoting environmental concerns is likely to diminish the priority given to the environment. Much has been written about the rise of environmental movements (Princen and Finger, 1994), their diversity, their political role, their methods (Eccleston, 1996; Thomas, 1996), and the issues they are concerned about (Potter, 1996). Despite their evident importance as a political phenomenon, there remains considerable uncertainty about their achievements (Potter and Taylor, 1996). They seek influence both at the level of policy and at the level of values. Arguably they fit into an EM perspective. This is certainly Mol's view suggesting that they "change their ideology, and expand their traditional strategy of keeping the environment on the public and political agendas towards participation in direct negotiation with economic agents and state representatives close to the centre of the decision-making process." (1995, p. 58). At national and international levels NGOs are being incorporated into the development of policy initiatives.

At the same time, environmental interests will seek other avenues of political influence. Although there is much evidence of co-operation between environmental movements and their major targets in government and business, there is also much conflict. In the UK, as elsewhere, local groups and NGOs, often operating in combination, are found campaigning against particular policies or trying to prevent specific developments. They are sometimes engaged in developing alternative ideas, strategies or lifestyles. Examples abound. Thus, in Britain groups block road proposals or prevent nuclear waste dumps; in Germany they try to shut down the nuclear industry; in Australia they try to protect wilderness and stop wood-chipping; in Nigeria they campaign against oil pollution in the delta; in Indonesia they oppose logging and try to establish social forestry. The evidence is that environmental movements are not simply incorporated into the EM negotiating process; they are also engaged in conflict over environmental management and policy. To the extent that they are arguing for alternative strategies for sustainable development their stance is antagonistic to EM with its presumptions of adjustment and adaptation rather than any more profound change.

In the case of Britain (and other western countries) while co-operative environmental management strategies within the general context of EM have a part to play in dealing with environmental problems they are unlikely to provide a universal solution. They are most likely to work where economic and environmental goals are compatible. Where major economic or social changes are necessary, as for example in transport policy, conflict over policy is likely to occur. And there are some industrial activities where no amount of ecological modernisation is likely to remove the basis for conflict. It is to one of these, the nuclear industry that I now turn.

11.4 The Nuclear Industry - negotiation or conflict?

Ecological modernisation and the nuclear industry
The nuclear industry has been long established in the UK. Although its origins were in the military sphere, with the opening of the Berkeley power station in 1965 it became the first country to develop a civil nuclear industry. Today the country has a complete nuclear cycle (apart from uranium mining) including fuel enrichment and fabrication, electricity generation (accounting for around 30% of output), and (with France and Russia) a civil (as well as a military) reprocessing industry. It also has waste management facilities. As with other industries there are grounds for suggesting that the nuclear industry has undergone a process of ecological modernisation.

Modernisation at the technological level
In many respects the recent history of nuclear power in Britain shows evidence of ecological modernisation. At the *technological* level there has been an increasing emphasis on risk reduction and environmental protection. This has reflected growing public concern about the hazards of the nuclear industry, heightened from time to time by major accidents such as Three Mile Island in 1979 and Chernobyl in 1986. In the UK itself there have been major alarms, notably a fire at Windscale (now Sellafield) in 1957, the accidental release of contaminated material on the coast at Sellafield in 1983 and recurrent anxieties about leukaemia clusters at nuclear sites and accidental releases of radioactivity into the environment. Sellafield has remained a focus of concern culminating in the long-running debate over the development and commissioning of the major new facility, the Thermal Oxide Reprocessing Plant which was finally given the go-ahead in 1995. More recently, there has been concern surrounding the Dounreay plant in northern Scotland where contamination has been discovered on the foreshore nearby seabed. The transfer of nuclear materials resulting from reprocessing operations has generated international concern and is likely to be an

issue of growing political significance (akin to the conflicts over the movement of plutonium and wastes between France and Germany and Japan).

The response of the nuclear industry, prompted by government and public opinion, has been to try to provide reassurance through increasingly tighter safety and security measures. Environmental protection has undoubtedly become a key concern in the further development of the nuclear industry. It has been a major issue at the long-running planning inquiries such as Sizewell B (at 340 days it remains the longest ever planning inquiry in the UK) Hinkley Point B and the Nirex repository at Sellafield. At Sizewell the safety issues took up nearly half (150 days) of the total time (O'Riordan et al., 1988). In the case of the Nirex repository, proving a satisfactory safety case will be the acid test on which the decision to go ahead will depend. Overall, the detailed and increasing attention to environmental protection in the nuclear industry presents a strong case for an industry adapting to the demands of ecological modernisation.

State and market
There is some evidence, too, that the nuclear industry increasingly conforms to the political model of ecological modernisation whereby the state provides a regulatory framework within which the industry is able to operate in relative freedom. The industry has recently been partially privatised with a holding company British Energy plc owning two operating subsidiaries, Scottish Nuclear Ltd. and Nuclear Electric Ltd. responsible for the Advanced Gas-cooled reactors (AGRs) and one Pressurised Water Reactor (PWR) in Scotland and England. The older Magnox reactors remain in a government owned company Magnox Electric plc and the fuel services and reprocessing complex of British Nuclear Fuels Ltd. (BNFL) centred on Sellafield in Cumbria also remains under government ownership. The disposal of intermediate and low level radioactive wastes (ILW and LLW) is the responsibility of UK Nirex Ltd. owned by the operating companies. But, overall, the industry has become less dependent on government control and been given greater freedom to compete in the market for electricity supply or to secure foreign contracts for fuel services, notably reprocessing where BNFL is in direct competition with the French reprocessing company COGEMA. The state's role is both regulatory and facilitative. Radiation standards are set by the National Radiological Protection Board and the authorisation, licensing and inspection of nuclear facilities comes under the aegis of the HM Inspectorate of Pollution (in the case of radioactive waste) and HM Nuclear Installations Inspectorate (NII)(in the case of nuclear plant). The state and its agent authorities provide the parameters of safety within which the nuclear companies are relatively free to make their own decisions. For instance, although the government is ultimately responsible for the development of a deep repository for nuclear waste, it is up to Nirex to select a site, apply for permission, develop a

safety case, and if successful at a planning inquiry, to operate the repository (rough future arrangements are now under discussion).

Openness and Participation

For long the nuclear industry was the archetype of secretive decision making. As Falk comments, "Since its inception the nuclear industry has been characterized by an exceptionally intimate relationship with the state" (1982, p. 116). The decision makers comprised the military, scientific advisers, government officials and politicians. Many of the decisions were taken without public debate, made in the name of energy security, national prestige, or the self-interest of the atomic institutions (Pringle and Spigelman, 1983, p. xi). In recent years there have been signs of penetration of this closed decision making. A number of factors have contributed to this. First, the industry's claims for safety were seriously undermined by major accidents especially at Chernobyl. Second, nuclear energy is costly and has suffered a commercial decline. As a result, thirdly, the focus has switched from the production of nuclear energy to the problem of cleaning up contaminated sites and managing nuclear wastes. And fourth, public opinion, influenced by environmental groups and the media has become increasingly hostile. The industry has become vulnerable. It has sought support through public information campaigns, consultation exercises and encouraged a culture of openness. Environmental groups, for long kept outside the decision making arena, are participating in the development of policy by incorporation into the policy making process. Local groups and national NGOs (notably Greenpeace and Friends of the Earth) are now routinely consulted, sometimes offered observer status at meetings though not, as yet, given membership of advisory bodies such as the Radioactive Waste Management Advisory Committee (RWMAC). In terms of participation the nuclear industry appears to fulfil the requirements of the ecological modernisation model.

Room for Doubt

Despite the evidence that the nuclear industry tends to conform to the precepts of ecological modernisation, a powerful contrary case can be established. In the first place the industry remains much more in the hands of the state than is true for most other industries. It is either owned by the state, heavily subsidised by the state or closely scrutinised by the state. Decision making remains, on the whole, elitist, dependent on a close relationship between government and industry. Thus, the kind of partnership arrangements between public and private reaching agreements through negotiation, is by definition precluded, certainly in the state-owned sector. Even in the privatised sector, the state necessarily keeps a tight control over environmental protection.

Although there has been greater openness and opportunity for public scrutiny of the industry's plans, proposals and research, a reputation for secrecy persists. This is reinforced by periodic revelations of problems which suggest at best a failure to provide information or, at worst, cover-up. For example, the contamination resulting from waste disposal operations at Dounreay (occurring in the mid-1960s) which had been monitored from 1979, was not disclosed until 1993 (RWMAC/COMARE, 1995). The Nirex policy of greater openness, including public information and peer review (Nirex, 1997; RWMAC, 1997) was found to be flawed when a leaked memorandum cast serious doubt on its claims that a safety case could be made for the Sellafield repository. The memo. conceded that Nirex might have to conclude that the site "is inherently not characterisable to the requisite level".

While environmental groups have been routinely consulted over the industry's proposals, there is little evidence of any close participation in the decision making processes. Indeed, the nuclear industry remains a major target of environmental movements. Conflict rather than consensus characterises the relationship between the industry and environmental movements. Of course, dialogue occurs (sometimes covertly) but environmental groups such as Greenpeace and Friends of the Earth have maintained the independence of action on which their ability to influence opinion depends. Their activities have not seemingly become marginalised nor have their activists become incorporated as part of a process of ecological modernisation.

Above all, it can be argued that the nuclear industry fails the central tenet of the ecological modernisation case. In the final analysis it is not susceptible to ecological modernisation since, while the risks associated with nuclear technology may be reduced, they cannot be eliminated. The statistical risk associated with new plants has been progressively reduced but, at the same time, the risks from ageing plants (notably in Eastern Europe) from radioactive wastes and from proliferation are widespread and persistent. From time to time, as Chernobyl has demonstrated, risk can be manifested in catastrophe. The least likely event can occur at any time, the worst case scenario is, ultimately, possible.

11.5. A Sustainable Future?

Limits to Progress

The analysis offered in this chapter suggests that ecological modernisation (of which negotiated solutions are a part) expresses the response of contemporary liberal capitalism to the environmental problem. It has emerged as an approach that is ideologically attuned to the dominant political dispensation. Therein lies both its strength and weakness. Its strength is its feasibility, its widespread support, its linkage of environmental conservation with economic growth. But, ecological modernisation will only be supported by business and the state up to the point where it impinges on economic performance. Environmental movements will criticise ecological modernisation for not going far enough. Consequently, ecological modernisation represents a compromise, a weak expression of sustainable development that encourages improvements in resource conservation and pollution control within a context of continuing growth but which cannot effect fundamental changes to patterns of production which, ultimately, may threaten survival. Ecological modernisation confines its attention mainly to the economic and technological spheres - its political analysis is thin and its social analysis non-existent. The emphasis on partnership relates primarily to the business/government nexus and in so far as other interests such as environmental movements are embraced at all, it is through incorporation or consultation rather than democratic participation. Again, the nuclear industry offers some verification of the point.

Charting the way forward
In looking ahead two things are needed. First, we need to examine the changes that may be necessary and, second, to consider their feasibility. Both require considerable effort and, at this stage, only an indication of the task is given. The kinds of political, social and economic changes that will be necessary to deal with the environmental question have already been hinted at in the preceding analysis. Their broad outlines may now be delineated in terms of three propositions.

1. Commitment to long term environmental management.
In the version of ecological modernisation discussed in this chapter, it is assumed that environmental protection can be delivered through the market together with an enabling state. The problem with this is that environmental sustainability requires long-term management, well beyond the short term horizons of markets or governments. To achieve this a stronger, more interventionist role by the state is needed providing consistent support to environmental objectives. This does not imply a centralised state planning system, not least because such a system is likely to be politically unacceptable. What it does suggest is a combination of two modes of decision making, market and planning that will encourage economic efficiency and that is consistent with the sustainable management of environmental resources. What is envisaged is the development of a system of environmental planning as outlined in the report of the UK Town and Country Planning Association (Blowers, 1993 and 1997b). In principle, environmental planning would have the following features:

a) It would take account of future uncertainty by a precautionary approach. This requires that action must be taken, 'where three are good grounds for judging that action taken promptly at comparatively low cost may avoid more costly damage later, or that irreversible effects may follow if action is delayed' (HMSO, 1990, p. 11). Of course, in some cases (e.g. radioactive waste) it may also be better to avoid precipitate action (such as deep disposal) if the consequences of such action is uncertain. This is a case of 'Don't just do something - stand there!' But, the principle of precaution still applies.

b) A system of environmental planning would reflect the integrated nature of environmental processes. Such processes are both trans-media (they are trans-mitted through different pathways of air, land and water) trans-sectoral (they cut across conventional sectoral boundaries of decision making) and they are trans-boundary (they are not confined by political boundaries). These features imply a policy response that is able to transcend the vertical and territorial boundaries of decision making.

c) Environmental planning implies a synoptic strategic approach. Environmental plans would be developed at various scales, from the local up to the global. They would be based on an analysis of the state of the environment and perceived trends and on a definition of the targets to be met and the indicators by which to measure achievement. Meeting the commitments of the plan will depend on a combination of market mechanisms and regulatory controls. The mixture will vary over time and space.

Such an approach may appear both vague and unrealistic. Interestingly, though, an embryonic form of environmental planning is emerging. There are many examples of state of the environment reports, at global (OECD, 1991), national (e.g. Australia, 1996; UK, HMSO, 1992; Netherlands, RIVM, 1989) and sub-national levels. The extent of the problems, the trends and the need to address them are already being documented. In some cases prototype environmental plans have been developed. At the international level the Commission for Sustainable Development at the UN and the international environmental conventions all indicate a concerted effort to respond to the environmental problems that beset the world. It is true that much of this effort remains either rhetorical or peripheral to more pressing political priorities. But, it exists and the signs are that it persists. The momentum established at the Rio Conference shows no signs of abatement. Of course, the political changes necessary to make it happen are fundamental. In the developed countries the capitalist market economy appears impregnable and support for the notion of intervention is weak to the point of invisibility. As Reade, commenting on the fate of planning in Britain, points out, "the creation of an effective system of environmental planning in Britain, like the achievement of other kinds of radical social change, depends on the emergence of a far more moderate political climate than presently exists in this country" (1997, p. 86). But, as the environmental crisis becomes more evident the possibility of applying constraint on the market should not be discounted. After all, when survival is at stake, public and private interests become aligned.

2. Commitment to social equality.

We have seen that one consequence of ecological modernisation may be the displacement of environmental problems and the creation of pollution havens elsewhere. Or, the extension of modernisation in developing countries may increase the burden on ecosystems and lead to environmental degradation. A similar pattern of uneven development also occurs within the developed countries with some areas benefiting while others experience the negative externalities of the modernisation process. The process of ecological modernisation is an extension of the process of modernisation but with a greener tint. It adds an environmental dimension to patterns of social inequality.

Inequality is a prevalent condition, exacerbated by the modernising force of the capitalist market economy. The drive to competition, efficiency, and comparative advantage inevitably results in winners and losers. The environment is a loser both ways. The need to sustain economic growth encourages richer countries to consume more resources (often imported from poorer countries) and to create more pollution (again, often in poorer countries). The poor, in order simply to survive, consume and degrade ever dwindling resources. No amount of ecological modernisation seems likely to reverse these basic (and well catalogued) trends.

Social equality is a necessary though not sufficient condition for sustainable development. Greater equality, under present conditions of production and consumption, is only likely to lead to greater exploitation of the environment. Greater equality must be combined with changes in production processes and life-styles. But, without greater equality there can be no co-operation or commitment to change. As Hugh Stretton put it two decades ago:

> "If a general principle is wanted it can only be some principle of equality; but a principle of equality will usually contradict any simple principle of either growth or conservation. Conservation is not worth having if it merely shifts hardships from rich to poor, or from later to now. Growth is not worth having if if merely speeds up the rate at which the rich can guzzle resources which the poor need both now and later" (1976, pp. 4-5).

Although by many standards of measurement the gap between rich and poor has tended to widen (certainly in the UK), paradoxically the issue of equality has diminished as a political priority. Yet, redistribution is a necessary measure if patterns of production and consumption are to be changed in order to secure environmental resources. The problem of social equality is linked to the issue of environmental management and state intervention. In Schnaiberg's words:

"We must first estimate a biospherically and geopolitically feasible, sustainable production level for the society. And then we must decide how to allocate the production options and the fruits of such production. Neither the state nor the market currently perform all these functions in any system - socialist or capitalist. Therefore, the intervention of the state is necessary to moderate the workings of market systems, though to varying degrees" (1980, p. 431).

To environmental management under conditions of social equality a third commitment must be added.

3. Commitment to Political Participation.

Ecological modernisation has been achieved through a synthesis of the state and the market in which the state plays a facilitating role for business to modernise its production processes, including those processes which contribute to environmental protection. The process of bargaining and compromise is not particularly new; indeed, it has been typical of the pragmatic British approach to environmental regulation. It is a system of decision making based on administrative discretion, self-regulation and compromise. Although solutions are negotiated, the interests of business and the environment may be incompatible. Consequently, "regulatory control is characterised by an ambivalence which has both political and moral dimensions" (Hawkins, 1984, pp. 8-9). 'Partnership' is a strictly limited idea. In the first place, the business interest tends to be predominant since business is only likely to undertake what will ultimately be good for business. Secondly, other interested parties are either offered some limited involvement (through consultation) or excluded altogether. The key decisions are still located in boardrooms or in government offices.

Participation in decision making implies a notion of partnership in which all have a share and from which all can perceive and draw benefit. Already the vigorous debates promoted by environmental movements are producing new ideas that run counter to the dominant materialist and consumerist values on which the 'treadmill of production' is predicated. Environmental movements have demonstrated an ability to influence both policy and values. A commitment to greater participation could lead to greater co-operation thereby reducing the conflict over the goals of environmental policy. It would increase the number of interests involved but would fall well short of fully democratic participation. Environmental movements, like business corporations, are not necessarily representative of the interests of the wider society.

Increasing participation means the redistribution of power; in the words of the Maastricht Treaty a process "in which decisions are taken as closely as possible to the citizen". True subsidiarity means "making decisions at the lowest level compatible with attaining required objectives". Power and responsibility become fused when those exercising power are responsible for the outcome of their actions. At present power is concentrated and much power is exercised by business corporation beyond the reach even of national democratic constraints. In a truly democratic society, power would be much more widely dispersed so that citizens would have greater responsibility for shaping their environment. Taken together, a system of environmental planning, greater social equality and wider participation could provide the basis on which a sustainable environment could be achieved.

Identifying the necessary commitments for a sustainable society is merely a preliminary task. The subsequent, more difficult task, is to envisage what such a society might look like, what institutional form it might take, what life-styles it suggests and what values it encourages. Even more problematic is to chart the way forward to the achievement of such a society. But, survival may well depend on a dedication to such a task now.

12. CONCLUDING REMARKS: THE SCOPE OF CO-OPERATIVE MANAGEMENT

PETER DRIESSEN

12.1 Introduction

The preceding chapters of this book revolve around a specific form of environmental governance, namely co-operative management. That model assigns a key role to collaborative relations between governments, mediating non-governmental organizations, and private interests. The mechanism for change lies in communication and dialogue, the results of which are laid down in voluntary agreements among the participants. Uncertainty and complexity, which are part and parcel of modern environmental issues, are not seen as obstacles. Rather, co-operative management takes these characteristics as a point of departure for governance. Uncertainty and complexity are made 'manageable' through collaboration between public and private parties. Co-operative management is essentially different from other forms of governance that are used in environmental policy - governance based on the models of regulatory control, market regulation, civil society, and self-regulation. In Chapter 1, Glasbergen points out that these models are associated with particular scientific disciplines; thus, each is fed by a different wellspring of rationality.

In practice, the quandary is not usually which one of these governance models to choose but how to put together the right mixture. Which models are actually put to use, to what extent, and in what proportions depends on the developmental stage of the environmental policy in a given country, but not alone. The selection also depends on the political and administrative culture, the nature of the environmental issues, and the constellation of special interests with a stake in these issues. In this final chapter, we do not elaborate on how this mix of governance models should be composed - and with good reason, since that question cannot be answered in a way that can be generalized to all situations. Instead, we offer some concluding remarks on the value of the model of co-operative management.

The chapters of this book elucidate the model of co-operative management by describing how it has been applied in diverse countries and relating those experiences from the perspective of various scientific disciplines. Those narratives

P. Glasbergen (ed.), Co-operative Environmental Governance, 251–267.
© 1998 Kluwer Academic Publishers. Printed in the Netherlands.

go into the ambitions and achievements of this policy approach but also beyond that level. The authors put the experiences they describe into a wider context, making some critical remarks on the course of the policy process under review. Their critical evaluation includes the outcomes of that policy process, the links to other governance models, and the political culture in which the process takes place. In the following discussion, we turn the spotlight once again on the characteristics of co-operative management as a new form of environmental governance. In particular, we examine three aspects that make this model of governance special: the name of the game, the key players, and the playing field. The game refers to the interactions between the public and private actors who are involved. From the foregoing chapters, it is clear that these interactions are steered by two social mechanisms, namely the 'learning' process and the process of 'exchange'. Here, we consider how those two social mechanisms can operate concurrently and under what conditions they function. In practice, 'fairness' turns out to play an important role in mediating these mechanisms. We go on to consider who can be involved in the game: the players. This question is closely related to the democratic caliber of the decision-making process. Several of the authors in this book have made a penetrating analysis of this topic. Then we turn our attention to the playing field, to assess how much leeway there really is to apply forms of co-operative management in a real-world situation. This topic is tied to the question posed earlier about the relation between this model and other forms of governance. Finally, we examine the results that co-operative environmental management could produce in practice, highlighting what we should definitely *not* expect of this form of governance.

12.2 The game: interaction through learning and exchanging

Voluntary co-operation
The model of co-operative environmental management is based on the idea that essential changes in society can only get off to a successful start if governments are assured of co-operation in their governance efforts. They have to be able to count on the institutionalized private parties who have a direct stake in the policy problem at hand. Thereby, environmental interests are not the only crucial elements in this approach. This model also revolves around the interests (mainly economic) of those parties who are compelled to adapt their behavior in accordance with the environmental objectives. In this manner, environmental objectives are related to economic aims. The co-operation that takes shape between governments and private parties should be grounded in voluntarism. No party can force another to collaborate. This does not mean, however, that participation is entirely non-committal. As the previous chapters illustrate, the

parties involved want to see the outcome of the consultation in the form of covenants, contracts, or agreements. Depending on the nature of the agreements, those formal documents may be invested with a certain degree of legal authority. Incidentally, many of the chapters in this volume - namely the contributions by Gebers, Basse, Bergmann et al., Seyad et al., Glasbergen, and Enevoldsen - depict a wide range of applications of such agreements in Europe. Not only do the countries apply these arrangements differently, but the significance attached to them varies from one country to the next.

Ernste (Chapter 3) sees co-operative management as a form of New Public Management, drawing upon innovation in the way government policy is organized. In this connection, Ernste draws a comparison with the development of Lean Production in industry. Both Lean Production and New Public Management are organizational concepts. The objective of both is to increase the efficiency and effectiveness of organizations by restructuring them and by activating human resources. Intervention of this kind has some concrete implications. One is that co-operative relationships are first forged between the diverse parties; only then are those ties institutionalized. Another implication is that the parties are encouraged to think and act creatively.

Both the government and private companies can derive some benefit from these co-operative relationships - and from the ensuing voluntary agreements. In Chapter 6, Basse points out some of the advantages. On the basis of their collaboration, companies can try to adapt the environmental objectives and the required measures to suit their own specific operating conditions. Furthermore, companies can capitalize on their agreement with the government by elaborating upon it in their public relations. On the government side, one advantage is the heightened environmental awareness among manufacturing companies. The government wants firms to acknowledge the environmental impact of their operational activities. They should also realize that the responsibility for research, development, and rehabilitation of the environment cannot be passed on to society at large.

Most of the interaction that takes place between the parties involved in co-operative management is through consultation and dialogue. By communicating with each other, the parties gain new insight into problems; from a different vantage point, they may see some new solutions. Moreover, communication stimulates the parties to come forward with creative ideas. They 'learn' from one another how to approach environmental problems from a different - and hopefully better - angle, which in turn motivates them to engage in co-operative action. On the other hand, the interaction between the parties can also take on the character of negotiation. This is only natural; in order to make things happen, people have to make specific agreements with each other. In that light, the model of co-operative environmental management is based on the assumption that two social mechanisms will come into play. These mechanisms - 'learning' through dialogue and 'exchange' through negotiation - can help the parties reach consensus on a joint approach to the problem they have to deal with. The essence of a learning process is to make uncertainty and complexity comprehensible and thereby manageable. The process of exchange is preeminently suited to making trade-offs between opposing social interests and demands, thereby increasing the likelihood that the policy will eventually be carried out. Finally, both processes make a positive contribution to the acceptance of the policy - at least, that is the expectation. Let us now consider how those two social mechanisms can be put to use and how they are related to one another.

Learning processes

At various places in this book, the authors mention that environmental issues are extremely complex. They also point out the high degree of uncertainty that surrounds these issues. The definition of an environmental problem should be understood as a social construct. Scientific and technical knowledge may, in fact, play a role in defining it, but such insight is certainly not of ultimate importance (Hannigan, 1995). Besides having some basis in fact, estimates of threats and risks are made on the basis of intersubjectivity. In addition, conflicts of interest between the actors play a role; those interests are related to the problem (or to its possible solutions). Indeed, various parties will try to bring their political norms, values, and preferences into the process of defining and agenda-setting. The environmental issue is thus constructed in a social interaction process (see also: Haila, Chapter 4).

When policy processes are perceived as learning processes, it is assumed that problems and their possible solutions are diversely interpreted by members of society. Each party with an interest in an environmental issue has his own standpoints and specific sources of knowledge. Any confrontation between these views and sources can generate new knowledge. It may also lead to new perspectives on problems, approaches, and solutions. The aim of this learning process is twofold. On the one hand, it is meant to explicate the 'social construction' of the environmental issue. That is, the social relevance of the issue is legitimated. On the other hand, it is meant to reinforce the social construction of the issue. In other words, the learning process can help connect the definition of the environmental issue to the insights and options for action that exist among key groups in society.

It may be clear that a learning process can only occur if stakeholders take part in the process of setting the agenda and defining the policy problems. Furthermore, the interested parties would have to take part in the preparation and implementation of the policy. At the same time, those involved should keep an open mind with regard to each other's standpoints and sources of knowledge. They should assume an adaptive stance. In other words, they should be receptive to each other's input. People have to be willing to adapt traditional attitudes and views in light of new insights. That is the only way to nurture interaction and communication and thereby to cultivate learning. The point of the learning process is not to determine which views and sources of knowledge are the best or the most reliable. Rather, the real question - taking the diverse standpoints, sources of knowledge, and uncertainties into account - is how to formulate a generally accepted definition of the problem and how to determine the socially desirable directions for any solutions to that problem. When the stakeholders take part in a process such as this, there will be broader support for the policy. The legitimacy of the policy will thus be greater than if government were to impose a policy unilaterally on a target group. In fact, the government plays a completely different role in this learning process than it does in more traditional forms of governance. As Ernste puts it in Chapter 3, "the only way the state can influence such a social learning process is not so much by taking a specific position itself and making this position plausible, but rather by acting as an initiator and moderator of a discourse between all involved parties."

Learning processes arise in the course of virtually all forms of policy. Persons in charge of carrying out policy and other parties with an interest in that policy can gain experience with the way the policy actually works through the methods of ex-post and ex-ante evaluations as well as through monitoring techniques. Practical experience is an important way to learn. Besides learning by doing, co-operative management is also - in fact, mainly - geared to learning through debate (see Van de Graaf et al., 1996, pp. 1-14). This approach is centered on the confrontation between the basic assumptions of those who have a stake in the policy. The debate lays bare their conceptual frameworks, their sources of knowledge, and their motivations. Once these issues are out in the open, the parties who have an interest in the policy may be inclined to change their standpoints. One of the main characteristics of a policy-oriented learning process is that it generates more knowledge and understanding. But other features are no less important. It also leads to a continual change in policy as well as to an improved way of dealing with problems as a result of new insights.

This is not the place to go into detail on the diverse theoretical approaches to policy-oriented learning. (The interested reader is referred to other publications, including Argyris and Schön, 1978; Friedmann, 1989; Bennet and Howlett, 192; Sabatier and Jenkins-Smyth, 1993; Glasbergen, 1996.) It is important to mention, however, that any form of co-operative environmental management should be organized as a learning process. There are three different reasons to do so, according to Vermeulen et al. (1998, p. 53

a) With respect to complex environmental issues, there will be no agreement among mutually dependent actors on the description of the problem and the route to be taken to address it.
b) The knowledge that is required to resolve environmental issues is spread over various actors within the policy circuit and outside of it.
c) The consequences of policy actions do have an effect on the target groups, but there is generally insufficient feedback from those target groups to the environmental policy.

These points have several implications for the basic conditions that must be present for learning processes to occur. First of all, public and private stakeholders must be enabled to bring their views and desires into the policy process. Secondly, they must be encouraged to jointly identify the policy objectives and to study the options for interweaving the objectives (Glasbergen, 1996, p. 191).

Negotiating processes

As stated above, co-operative environmental management is not only characterized by learning processes; it also calls for negotiating processes. This is discussed in Chapter 2 by Meadowcroft, who also cites Gray (1989) on this matter. Co-operative management is a means to reach a negotiated settlement. The point of negotiation is to get diverse public and private parties to an agree on how to tackle a problem. The approach should take the interests peculiar to each of the parties into account as much as possible. In the course of finding that approach, these parties should not only engage in dialogue with one another but should also enter into transactions with each other. Just as there are reasons to engage in learning processes, there are also specific reasons to start up negotiating processes in the framework of co-operative environmental management. Negotiating processes are necessary for the following reasons:

a) Present-day environmental issues are complex and require a combined application of policy instruments.
b) Both public and private parties have options (instruments) at their disposal with which they can help resolve a problem.They will generally have a different opinion on the best way to resolve the problem. Consequently, they will also differ with respect to the application of policy instruments.
c) The relationship among stakeholders is not hierarchical. Thus, they can act independently of one another.

Negotiations are intended as a means to find a balance between the diverse interests. In that respect, negotiation is different from both collaboration and conflict. Collaboration occurs when parties have compatible interests and objectives and are better able to achieve those objectives through concerted action. By definition, collaboration leads to a win-win situation. Conflict is the most likely strategy when parties have incompatible interests. Furthermore, it occurs when one of the parties believes there is more to be achieved by fighting than by talking. This form of interaction is usually accompanied by power plays and leads to win-lose situations. In negotiations, an effort is made to avoid a win-lose situation by jointly exploring the possibilities for a win-win outcome. In that sense, this effort may be called consensus-oriented negotiation. Nonetheless, every negotiating process will have collaboration as well as adversarial elements. Negotiations are accompanied by transactions. One major condition for successful negotiations is that the parties must have something to exchange. The exchange can refer to many different things. There are absolutely no restrictions, unless set by the parties themselves. The previous chapters show diverse instances of transactions among parties. One kind of transaction takes the form of postponing regulation in exchange for voluntary co-operation in an effort to reach environmental targets. At the other end of the spectrum is an instance of financial support in the acquisition and use of more modern technology in exchange for accepting more stringent environmental norms. In the course of the negotiating processes, all parties will be expected to made concessions. The negotiations are said to be successful when any concession made - a loss - is amply compensated by the benefits that are brought in by the compromise.

Conditions

It is difficult to distinguish learning processes from negotiating processes in the context of co-operative environmental management. It is possible to make an analytical distinction, though, as we have done in the above discussion. In practice, however, the two processes are intermingled and are not recognizable as separate processes. Nonetheless, there are subtle differences in emphasis. Learning processes emphasize the transfer of knowledge and seek better understanding. In contrast, negotiations are ultimately geared to an exchange of ideas on options for action. Both processes are highly relevant to the model of co-operative environmental management. When the learning processes turn out to be successful, the negotiation processes will be more likely to lead to consensus. At the same time, there may be some tension between learning and making trade-offs. The reason is that knowledge and information are used strategically in negotiations. That is, they are used in a way that is primarily beneficial to the party who is in possession of specific sources of knowledge. That position might frustrate the learning process. In order to do justice to learning processes within the framework of negotiations, the interactions between the parties should be subjected to certain conditions. These conditions are mainly concerned with fairness. In other words, the conditions should give all parties in the interaction process an equal chance - or as good a chance as possible. (It should be noted that we are using the concept of fairness in a different sense than that used by

Enevoldsen in Chapter 10. There, he refers to fairness in participation. We return to this point in the next section.) Following Webler[1] (1995, pp. 51, 59), we may list the following conditions (or rules of the game):[2]

a. Every discourse participant must have access to the knowledge needed to make validity claims and criticize others.
b. Every discourse participant must have an equal opportunity to make validity claims to comprehensibility, truth, normative rightness, and sincerity.
c. Every discourse participant must have an equal opportunity to challenge the comprehensibility, truth, rightness, or sincerity of validity claims made by others.
d. Judgement about conflicting validity claims must be made using the most reliable methodological techniques available.
e. Every discourse participant must have an equal opportunity to influence the choice of how the final determination of validity will be made and to determine discourse closure.

These conditions have to be general in nature. For each specific application of co-operative management, these conditions must be worked out into concrete rules that the parties will have to take into account. The parties should be in agreement on the rules of the game before they can move on to the stage of dialogue and negotiation about how to approach the environmental issue in question. Each form of co-operative management will thereby be identifiable by its pre-negotiating phase. That is the phase in which the parties try first of all to reach consensus on the procedural aspects of the dialogue that they want to enter into jointly (Driessen and Vermeulen, 1995).

12.3 The players: broadening the base for decision-making

In co-operative management, it is not only the game that has to be carefully designed. Careful thought should also be given to the question of who should take part. This question is related to the democratic nature of the decision-making process. According to Meadowcroft (in Chapter 2), the aim of co-operative management is to make the policy process more open in terms of its basis for

---

[1] The distinction that Webler makes between conditions for competence and conditions regarding fairness is not mutually exclusice on all points. Therefore, some of the conditions have been grouped together around the core concept of fairness.
[2] Similar criteria for fairness (with respect to the process and with respect to the opportunities for participation) may also be found in Susskind and Cruikshank (1987, pp. 21-25).

participation and its transparency to outsiders. A broader involvement and greater transparency may be seen as an improvement on - or in any event an important supplement to - the traditional democratic process for decision-making in the framework of democratically elected bodies. The broader basis that is thereby created has an added advantage. It allows the government to act with greater authority. This, in turn, inspires greater confidence in the government's ability to implement the policy, thereby assuring the continuity of that policy.

Two remarks should be made with respect to the statement that co-operative management leads to better democratic decision-making. First, we should consider who is involved in the policy process and how public and private parties can participate. The second remark concerns the relation between co-operative decision-making and the way decisions are reached in democratically elected bodies. These two points warrant a brief discussion, which follows.

The participants

The first question is, who can or may participate? In Chapter 4, Haila expresses a clear standpoint on this issue. "Deliberation should be open to everybody who is motivated to participate; by fostering a broader participatory basis, a greater variety of experiences are collected, and a greater variety of values are brought together." In Chapter 10, Enevoldsen formulates a similar standpoint. Inspired by the theories of John Rawls and Robert Dahl, he has advanced two criteria of fair participation.

1) "Any private party with an express, legitimate stake in the policy matter to be decided upon should have the opportunity to be included as an active participant in the political decision process."

2) "Included private parties must have adequate opportunities, and equal opportunities, to participate in all phases of the political decision process."

However, it is hard to put these propositions into practice. In co-operative management, the quality of democratic decision-making can come under pressure. This model of governance does not pertain to decision-making in representative bodies at the national, regional, or local level. Rather, it concerns decisions made in networks of public and private actors. These networks are built up on the basis of dependency relationships involving public and private actors who are confronted with a particular environmental issue. Such networks may be said to be a representation of interests. Participation in these networks should not be taken for granted, nor may it be construed as a right. An actor can only take part after all parties who might possibly be involved have reached agreement on allowing that participant into the network. Thus, the process of drawing parties into the consultation can also be selective, in the sense that certain interested parties may be deliberately excluded. In the previous chapters, the authors have brought up the fact that environmental organizations are not always involved in a process whereby the government and the private sector enter into a covenant to reduce the environmental load. Co-operative management thus does not automatically make the policy process more open, in the sense of encouraging broader participation and greater transparency. The parties involved have to deliberately pursue and organize that openness. Only when they make a concerted effort in that direction can this governance model exert a positive effect on the democratic caliber of decision-making.

Furthermore, the ideal of having broader participation in the decision-making process can also be constrained by a problem of a more practical nature. First of all, the participants should be competent to make a contribution to the policy process. They should know something about the policy problem. They should also possess the learning capacity and the argumentation skills that are indispensable to successful participation in a dialogue about complex environmental issues (see also: Webler, 1995, p. 59). In light of these conditions, parties with a potential interest - usually, the weaker or less powerful parties - may be forced to drop out of the process. Incidentally, this problem is not peculiar to co-operative management. It also arises in other forms of political participation. Secondly, the capacity to engage in co-operative management will decrease in proportion to the increase in the number of parties. This may seem paradoxical, as the collaboration among multiple public and private parties in co-operative management is actually considered to be a major advantage. Nonetheless, some environmental issues involve so many parties that co-operation would become difficult if participation were to be at a maximum.[3] Thus, not everyone who shows an interest can be admitted to the policy process. To do so would lead to unworkable situations. It might be possible to introduce degrees of participation, ranging from expressing one's opinion to being consulted or bearing part of the decision-making responsibility. In that way, it is in any event guaranteed that all those with a vested interest can bring their standpoints and desires into the decision-making process. With regard to direct participation - being a co-decision-maker - priority should be given to organized interest groups that demonstrate some professional qualities. Sometimes, however, these organized interest groups are themselves lacking in a democratic structure and culture with respect to decision-making. Meadowcroft has also raised the question (in Chapter 2) of who these organizations really represent. The crux of the matter is the relation between

---

[3] In the Netherlands, a form of co-operative management is employed in the field of regional planning. Those regional projects are concerned with a complex set of problems that are related to the socioeconomic structure, the quality of the physical environment, and the protection and development of nature areas. When an effort is made to tackle these problems in an integral way, it soon turns out that numerous public and private parties emerge on the scene, all claiming to have an interest in a project of this type (Glasbergen, 1995).

those who represent interest groups and those whom they purport to represent. Does the basis actually have a say in the decision-making process? Can the parties assume that the members will ultimately go along with the measures that have been agreed upon? One last question that arises is whether or not all of the interests that are at stake can actually be represented in the policy process. For those aspects that impinge directly upon the interests of businesses or the environment in which ordinary people live and work, interest groups may be expected to set themselves up. This will not be as likely for other environmental problems - ones that are less visible and more diffuse. For instance, who will defend the interests that are associated with the greenhouse effect, the depletion of the ozone layer, or the insidious degradation of nature areas? How can these interests be drawn into the negotiation processes that are mainly taking place at the level of the government and the private sector?

Relation to formal decision-making

Co-operative management involves another kind of risk. That risk is embedded in the relation to the existing process of democratic decision-making. This relationship creates a major dilemma for the government. Co-operative management puts the government in a relatively new position. Traditionally, it governed in a hierarchical fashion, which entails democratic control by previously established legal regulations. The current situation calls for consultation and negotiation with other tiers of government and private parties, whereby the rules of the game differ in each case and other organizational forms are applied. When the parties come to agreement with each other, their consensus will help resolve the government's legitimation problem. The reason is that there will be broader support for government policy. Nonetheless, it remains to be seen how much influence democratically elected bodies can exert on decision-making processes. When they are completely excluded from the process, the government will get a reputation for conducting its political activities in an exclusionary manner. The problem is that the mandate of democratic decision-making in elected bodies - and the ensuing political deliberation - is thereby undermined. If the government does make room for the traditional democratic process, however, it might possibly run into problems as a negotiating partner. The reason is that the government cannot take any definitive standpoints. By its very nature, democratic decision-making is capable of undermining positions taken by governmental representatives.

In the Netherlands, a solution to this dilemma has been found. The government has laid down a regulation that stipulates - among other things - how the parliament is involved in covenants to which the national government is party. One of the indications is that obligations are only entered into on condition that parliament will agree to the terms. Furthermore, the regulation stipulates that parliament has to be heard if a covenant initiates a new policy, if the covenant touches upon topics that are politically sensitive, and when the covenant implies major financial commitments on the part of the government (Ministerie van Algemene Zaken, 1995).

Conditions
The ultimate goal of co-operative management is to enhance the legitimacy, the effectiveness, and the flexibility of the policy process. To that end, there is an intensive exchange of information among governments, private parties, and citizens. In addition, non-governmental parties are given the opportunity to influence policy as it is taking shape. Furthermore, those same parties are made partially responsible for the implementation of the policy. This broader scope for participation can lead to positive side effects on the democratic caliber of the decision-making process. In practice, however, this is easier said than done, as the previous chapters demonstrate. The problems arise mainly because the experience gained thus far with this new model of governance has been modest. Problems also arise because the process of incorporating this new model into existing political and administrative cultures is proceeding slowly in diverse European countries. The most powerful parties must not be allowed to dominate the policy process, however. This is a real danger; if the powerful players do get the upper hand, co-operative management will tend to reinforce the existing power structures rather than break them down (see Blowers, Chapter 11). This model has greater potential with respect to the democratic caliber of the decision-making than other models of governance, on condition that a few rules are taken into account. In any case, the rules would have to touch upon the following matters:

a. The rules should indicate which interests are at stake in selecting an approach to a given environmental issue.
b. The rules should indicate which parties represent these interests and which parties should be able to take part in the dialogue on the most desirable approach.
c. The rules should indicate how and at which point in time democratically elected bodies can have some input in the decision-making process.
d. The rules should indicate how civil organizations, companies, and those tiers of government that are not involved in the negotiating process can bring their standpoints and desires into the policy process.

With respect to the last of these four points, the European Commission has ruled that, before an environmental covenant can be closed, the interested parties must have the opportunity to comment on the draft text. "In addition to those actually negotiating the agreement, all relevant business associations or companies concerned, environmental protection groups, local or other public authorities concerned should therefore be appropriately informed and comments should be taken into consideration in the final negotiation of the agreement" (European Commission, 1996).

12.4 The playing field: arena for successful interaction

Co-operative environmental management can be applied in diverse situations. For instance, this form of governance can be applied when new environmental issues present themselves and governance by other means - by regulation, for example - is not feasible in the short term. Co-operative management offers an opportunity - through voluntary agreements with target groups - to take quick action in anticipation of a more hierarchical approach. In this case, there is a large amount of room to maneuver. The definition of the problems, the objectives, and the measures will be subject to dialogue and negotiation.

This form of governance can also be applied when the nature of the problem is known and the objectives of the policy have already been set but when the target groups have not yet reached agreement on the way in which those objectives should be achieved. In that case, there is less room to maneuver. The negotiations will only deal with the implementation process.

Finally, co-operative management can be applied in combination with other instruments of governance (legal or economic instruments). That happens in circumstances whereby those other instruments of governance prove to be not as successful as expected. Then, government authorities and target groups have to make additional arrangements to get the process of policy implementation moving again. In that case, there will be little leeway for consultation and negotiations. In principle, there is limited scope because the main contours of the terrain have already been staked out with other governance instruments. Nonetheless, this does not exclude the possibility that private parties who are forced to adapt their behavior as a result of the environmental measures will try to turn prevailing opinion in their favor. At least they will try to raise the issue in the process of consultation and negotiations, calling for a review of the policy objectives and the measures designed to implement them. In that event, it would be to the advantage of government parties to keep the scope for action as narrow as possible. The reverse would be true of other parties, who would want to continually widen their room to maneuver. This tension will build up in the course of any negotiating process.

Even when the playing field seems to be as big as it will get, there will always be legal regulations that have to be taken into account. Therefore, a crucial question may be raised in a general sense: what is the relation between this consensual approach to governance and the more hierarchical means of governance that are based on prescription, prohibitions, and strict norms? In other words, to what extent does law restrict the amount of leeway that forms of co-operative management have? In answering this question, three remarks are in order.

First of all, Glasbergen makes it abundantly clear in Chapter 7 that co-operative management - as well as the ensuing voluntary agreement or covenant - must not contravene the system of administrative law. Covenants can be used proactively in anticipation of regulatory measures yet to be passed. Covenants can also be used to give added impetus to the implementation of measures required by existing law. However, covenants cannot be used to create situations that would supersede existing legislation. For instance, a covenant cannot make it possible to refrain from issuing permits while such a permitting procedure is required under current law. It is conceivable that parties might conclude - as the outcome of negotiations - that solutions for a particular environmental issue are not possible within the framework of existing legislation and that the government would therefore see fit to change the law.

The second point that warrants mention is the position of third parties, particularly those who have not been included in the negotiations but who do feel the consequences of what the public and private parties have agreed upon. For instance, the interests of local residents may be affected by a voluntary agreement between the government and companies with respect to abating nuisance - namely noise and odor - and reducing the amount of waste they generate. When these agreements are subsequently translated into checklists for granting permits to the companies in question, citizens have an opportunity to challenge the agreement. If it comes to that, the administrative court will always consider whether sufficient weight had been given to local interests when government authorities and the business community hammered out an agreement. It is precisely the way formal decision-making processes are affected by agreements reached through co-operative management that provides some quality assurance. The impact on decision-making can help ensure that the procedures leading to such agreements will be followed as conscientiously as they should be. In that sense, a link between consensual governance and hierarchical governance may mean a step forward.

Third, the threat of regulation can be used as a stick to back up negotiations with private parties, as Seyad et al. point out in Chapter 6. In that way, private parties can be prevented from using delaying tactics or from forcing the government to make major concessions. In the event that the parties do not live up to their agreements, legislation may offer an alternative course of action. There is a disadvantage, though. Because it may take years to get a law passed, the threat of legislation is not always credible (unless a draft law has already been prepared).

Finally, one wonders how free government authorities and private parties really are to shape co-operative management according to their own insights. Earlier in this chapter, it was pointed out that complete freedom can have unforeseen effects. The most powerful parties might just use this form of governance to strengthen their influence on policy. Yet that runs counter to the principles of co-operative management, namely broad participation, transparency, and fairness. Rules could also be established for co-operative management. The above sections of this chapter made explicit reference to the rules of the game rather than to juridical rules. Co-operative management is based on *voluntary* collaboration between public and private parties. Most benefits come from this voluntary form of co-operation, because the voluntary aspect enhances the policy's capacity for governance and problem resolution - at least that is the assumption. In that connection, it is also important that the parties themselves can set the rules for the process of consultation and negotiation. Any attempt to make the process more efficient through institutionalization and formalization would undoubtedly make it less effective. Still, to ensure that the decision-making process will retain its open character, the government would merely have to establish its right to exert an influence by way of a general regulation or order. If third parties were to challenge the open character of co-operative management, they would always be able to fall back on this general order in any appeal they might lodge. For instance, the Netherlands has a General Administrative Law that includes regulations on conscientiousness, weighing of interests, and motivation of decisions. These criteria apply to general governmental decisions. Yet they could also be declared applicable to voluntary agreements that are made between government authorities and private parties.

12.5 Final remarks

The mission of environmental policy is to address societal activities that pose a threat to the quality of the physical environment. Any such corrective measures entail subjecting current activities to conditions. Private companies that would be compelled by the environmental policy to adapt their operational activities will not necessarily do so for economic reasons. The central question in environmental policy thus becomes how to develop a policy that not only lends itself to implementation but can also count on acceptance and even active support from those who are affected by the policy. Indeed, acceptance and support are crucial to an effective policy.

Various forms of governance can be applied in environmental policy, as Glasbergen points out in Chapter 1. Each of these forms of governance has its advantages and its disadvantages. The chance that it will lead to a successful resolution of a problem is limited in each case. It is generally expected that the co-operative management model will increase the chances of societal change. The reason is that this model calls for a concerted effort by public and private parties. Through intensive interaction - whereby learning and exchange are central, as argued above - an effort is made to reach consensus on the best way to deal with the problem. By also making private parties responsible for the policy, the chances that it will actually be implemented are increased. Moreover, it will be possible to muster the maximum amount of support for the policy in this way.

In light of the contributions to this book, it is clear that co-operative management is not an independent policy instrument. Rather, it is always applied in combination with other instruments. It is precisely in the relation with other instruments of governance that co-operative management can demonstrate its added value.

We should not have unrealistic expectations of the outcome of the consensus that is reached between public and private parties under the model of co-operative management. The reason is that consensus does not imply a breakthrough in the conflict of interest that exists between the parties. Even though the interests continue to differ, the parties may be expected to deal with them better. Through learning processes, they look for common points of departure for defining the problem and finding a solution. Trade-offs are made in an attempt to harmonize the actions of the parties in such a way that the environmental issue can be successfully addressed and the interests of the diverse parties can be accommodated, at least to a large degree.

Consensus should not be seen as a given. It has to be continually reconfirmed or reformulated in the course of the implementation of the policy. The societal and economic context in which the policy is conducted should always be assumed to be subject to change. The foundation on which the consensus was built can thus disintegrate. Co-operative management should therefore also be highly flexible. Furthermore, co-operative management will not lead to a reduction in conflicts between public and private parties. Indeed, this method of governance was not designed to prevent or resolve conflicts but rather to make conflicts more manageable. Conflicts are known for their negative effects - such as an increased chance of disintegration, stress, and distortion of reality - but they also have a positive impact. They are a source of renewal and change. Conflicts may make people more aware of the interests they are dealing with. Moreover, conflicts have a stimulating effect on motivation and creativity (Ritsema van Eck and Huguenin, 1993). These positive effects are put to good use in co-operative management. It is even conceivable that the deliberate provocation of conflicts between governments and private parties could offer a strategy for getting an interaction process off the ground, a process whereby a concerted effort is made to tackle a given problem.

In view of the preceding chapters of this book, it is clear that we should not expect too much from co-operative management. The solutions usually remain within the bounds of what is economically possible and feasible. That should come as no surprise. One cannot realistically expect private parties to voluntarily go along with measures that severely debilitate their economic position. In Chapter 11, Blowers rightly asks if this method offers sufficient grounds for sustainable development. Co-operative management is not only directed toward intensive collaboration between the government and the business community. It is also geared to technological innovations that can minimize negative environmental effects. Moreover, its success depends upon the performance of the liberal market economy. There is some doubt that this strategy can provide a basis for dealing with issues at a global scale, such as climate change and the environmental problems associated with underdevelopment. On the other hand, from an international perspective, there seems to be no other strategy available. The only way ahead seems to be through consultation and negotiation, which can prepare the ground for efforts to deal with these major issues.

BIBLIOGRAPHY

Aalders, M.V.C. (1993) Het milieuconvenant wordt 'salonfähig'. Convenanten nieuwe stijl in het milieubeleid, in J.Th.A. Bressers et al. (red.), *Beleidsinstrumenten bestuurskundig beschouwd*, Van Gorcum, Assen, pp. 75-92

Algemene Rekenkamer (1995) *Convenanten van het Rijk met bedrijven en instellingen*, Sdu Uitgeverij, Den Haag

Andersen, M.S. (1994) *Governance by Green Taxes*, Manchester University Press, Manchester

Antoni, C.H. (1990) *Qualitätszirkel als Modell partizipativer Gruppenarbeit. Analyse der Möglichkeiten und Grenzen aus der Sicht der betroffener Mitarbeiter*, Huber, Berne

Argyris, C. and D. Schön (1978) *Organizational Learning: A Theory of Action Perspective*, Addison Wesley, Reading

Artois, J. (1993) *Verpakking en afval in Vlaanderen: een integrale aanpak*

Åström, Sven-E. (1978) *Natur och byte. Ekologiska synpunkter på Finlands ekonomiska historia*, Helsingfors, Söderströms & Co.

Aulinger, A. (1996) *(Ko-)Operation Ökologie. Kooperation im Rahmen ökologischer Unternehmenspolitik*, Metropolis, Marburg

Autorecycling 1996 (1) (1996) *Herausgegeben von der Arbeitsgemeinschaft Deutscher Autorecyclingbetriebe GmbH*, Köln

Bachrach, P. and M. Baratz (1970) *Power and Poverty*, Oxford University Press

Baeke S., A. Seyad, M. De Clercq (*1997) Packaging Waste in Belgium*, draft

Barns, I. (1995) Environment, democracy and community, *Environmental Politics*, vol. 4, no. 4, pp. 101-133 (Special Issue: Ecology and Democracy)

Bartlett, R.V. (1986) Ecological rationality: reason and environmental policy, *Environmental Ethics*, 8, pp. 221-239.

Basse, E.M. (1997) Environmental contracts - as tools to implement international obligations, in Basse (ed.), *Environmental law: from international to national law*, Sweden Maxwell & GadJura

Bastmeijer, K. (1994) *Provisional code of conduct for concluding environmental covenants*, VROM, Den Haag

Bastmeijer, K. (1996) *The covenant as an instrument of environmental policy in the Netherlands: a case study for the OECD*, VROM, Den Haag

Baumol, D.W. and W.E. Oates (1988) *The theory of Environmental Policy*, Cambridge University Press

BDI - Bundesverband der Deutschen Industrie e.V. (1996) *Aktualisierte Erklärung der deutschen Wirtschaft zur Klimavorsorge*, Köln

Beck, U. (1986) *Risikogesellschaft. Auf dem Weg in eine andere Moderne* (English translation: Risk Society: Towards a New Modernity, Sage, London 1992), Suhrkamp, Frankfurt am Main

Beck, U. and E. Beck-Gernsheim (eds.) (1994) *Risikante Freiheiten. Individualiserung in modernen Gesellschaft*, Suhrkamp, Frankfurt am Maim

Beck, U. (1995) *Ecological Politics in an Age of Risk*, Polity Press, Cambridge

Beck. U. (1996) Risk Society and the provident state, in S. Lash et al. *Risk, Environment and Modernity: Towards a New Ecology*, Sage, London, pp. 27 - 43

Bennett, C. and M. Howlett (1992) The lessons of learning: reconciling theories of policy learning and policy change, *Policy Sciences*, 25, pp. 275-192

Benzler, G. and K. Löbbe (1995) Rücknahme von Altautos - Eine kritische Würdigung des Konzepte, *RWI-Mitteilungen*, Jg. 45, 2, pp. 141-162

Beyer, Wolfgang (1986) *Der öffentlich-rechtliche Vertrag, informales Handeln der Behörden und Selbstverpflichtungen Privater als Instrument des Umweltschutzes*, Dissertation Universität Köln

Bichsel, A. (1996) *NGO's as agents of public accountability and democratization in intergovermental forums*, pp. 234-255 in Lafferty and Meadowcroft (eds.), op.cit.

Biekart, J.W. (1993) *De uitvoering van het NMP door de industrie*, Stichting Natuur en Milieu

Biekart, J.W. (1994) *De basismetaalindustrie en het doelgroepenbeleid industrie. Analyse van proces en resultaten op weg naar 2000*, Stichting Natuur en Milieu

Biekart, J.W. (1995) Environmental covenants between government and industry: A Dutch NGOs experience, *Reciel*, 4, 2, pp. 141-149

Biekart, J.W. (1996) Green groups: new instruments as panacea or pandora's box. A: negotiated agreements, in J. Golub, *New instruments for environmental protection in the European Union*, Routledge (forthcoming)

Bingham, G. (1986) *Resolving environmental disputes*, Conservation Foundation, Washington D.C.

Bjerregaard, R. (1995) *Frivillige aftaler kræver offentlig indsigt og kontrol* [Voluntary agreements require public access and control], Jyllandsposten

Bleicher, K. (1992, 2nd edn.) *Das Konzept Integriertes Management*, Campus, Frankfurt am Main

Blowers, A. (1984) *Something in the Air: Corporate Power and the Environment*, Harper and Row, London

Blowers A. (ed.) (1993) *Planning for a Sustainable Environment*, Earthscan, London

Blowers, A. and P. Glasbergen (1995) The Search for Sustainable Development, in P. Glasbergen and A. Blowers (eds.) *Environmental Policy in an International Context: Perspectives*, Arnold, London, pp. 163 - 183

Blowers, A. (1997a) Environmental policy-Ecological Modernisation or the Risk Society?, *Urban Studies* (forthcoming)

Blowers, A, (1997b) Society and Sustainability: the Context of Change for Planning, in A. Blowers and B. Evans (eds.) *Town Planning into the 21st. Century*, Routledge, London

BMU - Bundesministerium für Umwelt, Naturschutz und Reaktorsicherheit (1995) *Staatssekretär Jauck fordert mehr Kooperation von Staat und Wirtschaft im Umweltschutz*, BMU-Pressemitteilung 117/95, Bonn

BMV - Bundesminister für Verkehr (1995) *Statement des Bundesministers für Verkehr anläßlich der Präsentation des Opel Corsa Eco 3 am 7.9.1996 in Bonn*, Fs. Nr.: 0340 070995 1430.sch

Bocken H., J. Bouckaert et al. (1994) *Milieubeleidsovereenkomsten, Wetenschappelijk verslag Milieu en Natuurrapport Vlaanderen 1994*, Gent

Boulan, R.P. (1994) Beschouwing van het doelgroepenbeleid in de basismetaalindustrie, RIZA

Brandstoffen gebruikt voor de produktie van elektriciteit in België, BFE Statistisch Jaarboek 1992, 1993 en 1994

Braudel, F. (1972) *The Mediterranean and the mediterranean World in the Age of Philip II* (vol. 1), William Collins Sons & co., London

Braudel, F. (1982) *Civilization & Capitalism 15th - 18th Century*, vol 2. The Wheels of Commerce, William Collins Sons & Co., London

Braudel, F. (1984) *Civilization & Capitalism 15th - 18th Century*, vol 3. The Perspective of the World, William Collins Sons & Co., London

Brenck, A. et al. (1996) *Zwischen Entsorgungsnotstand und Überkapazitäten - Wege zur effizienten Organisation einer umweltverträglichen Abfallwirtschaft*. Schlußbericht zur Studie "Optimierung des Verbrauchs knapper Deponiekapazitäten" für den Bundesminister für Wirtschaft, Zusammenfassung der Ergebnisse, Berlin

Brennan, A. (1992) *Moral pluralism and the environment*, Environmental Values, 1, pp. 15-33

Bressers, J.Th.A. and P-J. Klok (1996) Ontwikkelingen in het Nederlandse milieubeleid: doelrationaliteit of cultuurverschuiving?, in *Dynamiek in beleid en beleidswetenschap*, Jubileumnummer Beleidswetenschap, pp. 445-460

Brockmann, K.L., H. Koschel, I. Kühn, K. Rennings (1995) *Ordnungspolitische Grundfragen einer Politik für eine nachhaltige, zukunftsverträgliche Entwicklung. Expertise im Auftrag des Bundesministeriums für Wirtschaft*, Zentrum für Europäische Wirtschaftsforschung, Mannheim

Broek, J.H.G. van den (1993) Covenant and permit in the dutch target group consultation. The role of an environmental convenant in the granting of an environmental permit, pp. 33-34, in van Dunné, op.cit.

Brohm, W. (1992) *Rechtsgrundsätze für normersetzende Absprachen - Zur Situation von Rechtsverordnungen, Satzungen und Gesetzen durch Kooperatives Verwaltungshandeln*, DÖV, pp. 1025 - 1035

Brohm, W. (1994) *Rechtsstaatliche Vorgaben für informelles Verwaltungshandeln*, DVBl., pp. 133 - 139 Bundesregierung (1996a) Erklärung der deutschen Wirtschaft zur Klimavorsorge, Pressemitteilung Nr. 118/96, Bonn

Bungard, W. (1984) Das Training von QZ-Moderatoren, in M.E. Strombach (ed.) *Qualitätszirkel und Kleingruppenarbeit als praktische Organisationsentwicklung*, pp. 432-456, Kommentator Verlag, Frankfurt

Bungard, W. and M.E. Strombach (1986) Qualitätszirkel auf dem Prüfstand, *Harvard Manager*, vol. 4, pp. 40-45

Buuren, P.J.J. van (1993) Environmental covenants possibilities and impossibilities: an administrative lawyer's view, pp. 49-55 in J.M. van Dunné (ed.), op.cit.

Caldwell, L.K. (1970) *Environment: A Challenge to Modern Society*, Garden City, N.Y., Natural history Press

Carew-Reid, J. et al. (1994) *Strategies for National Sustainable Development*, Earthscan in association with IIED and IUCN, London

Carley, M. and I. Christie (1992) *Managing Sustainable Development*, Earthscan, London

Carlisle, J. and R.C. Parker (1989) *Beyond Negotiation. Redeeming Customer-Supplier Relationships*, Wiley, Chichester

Cawson, A. (1986) *Corporatism and Political Theory*, Basil Blackwell, Oxford

Christoff, M. (1996) Ecological modernisation, Ecological modernities, *Environmental Politics*, vol. 5, no. 3, pp. 476 - 500

Claus, F. and P.M. Wiedemann, (eds.) (1994) *Umweltkonflikte. Vermittlungsverfahren zu ihrer Lösung, Blottner*, Taunusstein

Commissie voor toetsing van wetgevingsprojecten (1992) *Convenanten*, Den Haag

Committee on Medical Aspects of Radiation in the Environment/Radioactive Waste Management Advisory Committee (COMARE/RWMAC) (1995) *Report on Potential Health Effects and Possible Sources of Radioactive Particles Found in the Vicinity of the Dounreay Nuclear Establishment*, HMSO, London

Cook, E. (1996) Marking a Milestone in Ozone Protection: Learning from the CFC Phase-out, in *World Resources Institute. Issues and Ideas*, January 1996, Washington

Costain, W. and J. Lester (1995) The evolution of environmentalism, in J. Lester (ed.), *Environmental Politics and Policy*, Duke University Press, Durham

Crenson, M. (1971) *The Un-Politics of Air Pollution*, The Johns Hopkins Press, Baltimore

Cronon, W. (1985) *Changes in the Land. Indians, Colonistis, and the Ecology of New England*, Hill and Wang, New York

Cronon, W. (1991) *Nature's Metropolis. Chicago and the Great West*, W.W.Norton & Co., New York

Cuijpers C., S. Proost et al. (1994) *Energie en Milieu in het Federaal België: Historiek en Toekomstverkenning*, Leuven

Dahl, R.A. (1956) *A Preface to Democratic Theory*, University of Chicago Press, Chicago

Dahl, R.A. (1971) *Polyarchy, participation and oppositon*, Yale University Press, New Haven

Dahl, R.A. (1989) *Democracy and its critics*, Yale University Press, New Haven

De Clercq, M. (1996a) The political Economy of Green Taxes: The Belgian Experience, *Environmental and Resource Economics*, no. 8, pp. 273-291

De Clercq, M. (1996b) The Dynamics of Interaction between Industry and Politics. The Introduction of Ecotaxes in Belgium, *Business Strategy and the Environment*, vol. 5, no. 3, pp. 207-215

Denhardt, R.B. (1981) Toward a critical theory of public organisation, in *Public Administration Review*, vol. 43, pp. 174-179

Denhardt, R.B. (1993, 2nd edn.) *Theories of Public Organisation*, Hartcourt Brace College Publishers, Fort Worth

Deppe, J. (1990, 2nd edn.) *Quality circle und Lernstatt: ein integrativer Ansatz*, Gabler, Wiesbaden

Dietz, F. (1994) Natuurlijke hulpbronnen in de neoklassieke theorie, in F. Dietz, W. Hafkamp, J. van der Straaten, *Basisboek Milieu-Economie*, Boom, pp. 38-58

Doherty, B., and M. de Geus (eds.) (1996) *Democracy & Green Political Thought. Sustainability, Rights and Citizenship*, Routledge, London

Donner-Amnell, J. (1991) Metsäteollisuus yhteiskunnallisena kysymyksenä Suomessa, in I. Massa and R. Sairinen (eds.), Ympäristökysymys. Ympäristöuhkien haaste yhteiskunnalle (pp. 265-306), Gaudeamus, Helsinki

Driessen, P. and W. Vermeulen (1995) Network management in perspective, pp. 155-178 in P. Glasbergen (ed.), op.cit.

Dryzek, J.S. (1983) Ecological rationality, *International Journal of Environmental Studies*, 21, pp. 5-10

Dryzek, J.S. (1987) *Rational Ecology: Environmnent and Political Economy*, Basil Blackwell, Oxford

Dryzek, J.S. (1990) *Discursive Democracy. Politics, Policy, and Political Science*, Cambridge University Press

Dryzek, J.S. (1995) Towards an ecological modernity, *Policy Sciences*, vol. 28, pp. 231 - 242

Dryzek, J.S. (1995) Political and ecological Communication, *Environmental Politics*, vol.4, no.4, pp. 13-30 (Special Issue: Ecology and Democracy)

Dryzek, J.S. (1996) Strategies of ecological democratization, pp. 108-123, in Lafferty and Meadowcroft (eds.), op.cit.

Dryzek, J.S. (1997) *The Politics of the Earth: Environmental Discourses*, Oxford University Press, Oxford

Dunlap, R. (1995) Public opinion and environmental policy, in J. Lester (ed.), *Environmental Politics and Policy* (2nd. edition), Duke University Press, Durham

Dunn, W. N. and B. Fozouni (1976) *Towards a Critical Administrative Theory*, Sage, Beverly Hills

Dunné, J.M van (ed.) (1993) *Environmental Contracts and Covenants, New Instruments for a Realistic Environmental Policy?*, Koninklijke Vermande, Lelystad

Dyke, C. (1988) *The Evolutionary Dynamics of Complex Systems. A Study in Biosocial Complexity*, Oxford University Press, Oxford

Dyke, C. (1994) The world around us and how we make it: human ecology as human artefact, *Advances in Human Ecology*, 3, pp. 1-22

Dyke, C. (1997) The heuristics of ecological interactions, *Advances in Human Ecology*, 6, pp. 49-74

Eckersley, R. (1995) Liberal democracy and the rights of nature: the struggle for inclusion, *Environmental Politics*, vol.4, no.4, pp. 169-198 (Special Issue: Ecology and Democracy)

Eckersley, R. (1995) *Markets, the State and the Environment*, Macmillan, London

Eden, S. (1996) Problems of Environmental Consensus: Packaging Regulation and Self-Regulation in the UK. Paper presented on the Sixth Annual Greening of Industry Network Conference, November 1996, Heidelberg

Eijlander, Ph., P.C. Gilhuis en J.A.F. Peters (1993) *Overheid en zelfregulering. Alibi voor vrijblijvendheid of prikkel tot actie?* W.E.J. Tjeenk Willink

Electrabel, Evolutie van de SO_2- en NO_x- emissies door de klassieke electriciteitscentrales in België

Endress, R. (1991) *Strategie und Taktik der Kooperation. Grundlagen der zwischen- und innerbetrieblichen Zusammenrabeit*, Schmidt, Berlin

Enquete-Kommission Schutz des Menschen und der Umwelt des Deutschen Bundestages (Hrsg.) (1994) *Die Industriegesellschaft gestalten - Perspektiven für einen nachhaltigen Umgang mit Stoff- und Materialströmen*, Bonn

Ernste, H. and S. Baumann (1995) Soziale Dilemmata und Umwelterziehung, in Greenpeace (eds.) *Neue Wege in der Umweltbildung. Beiträge zu einem handlungsorientierten und sozialen Lernen*, AOL-Verlag, Verlag die Werkstatt, pp. 136-152

Ernste, H. and M.J. Sirol (1995) Mit einer Schlankheitskur zur Umweltqualität, in *Das Magazin QUALITÄT*. vol. 30, no. 11-12, pp. 32-35

Ernste, H. (1995a) Trust, communicative rationality and the anthropological foundation of the man-environment-relationship, in P.P. Wright et al. (eds), *Human Ecology: Progress Through Integrative Perspectives*, pp. 5-13, Society for Human Ecology, Bar Harbor

Ernste, H. (1995b) *Kooperative Netzwerke und die Wiederentdeckung der Region*, Paper presented at the general colloquium of the Department of Geography, University of Berne, November 11, 1995

Ernste, H. (1996) Kommunikative Rationalität und umweltverantwortliches Handeln, in Kaufmann-Hayoz, A. Di Giulio (eds.) *Umweltproblem Mensch? Humanwissenschaftliche Zugänge zu umweltverantwortlichem Handeln*, pp. 197-216, Haupt, Berne

Ernste, H. (forthcoming 1997) *Über den Vernünftigen Umgang mit unserer Umwelt*, Habilitation thesis at the Faculty for Environmental-Natural-Sciences, ETH Zurich

European Commission (1996) *Communication to the Council and European Parliament on Environmental Agreements*, COM(96)561, 27 November

Falk, J. (1982) *Global Fission: The Battle over Nuclear Power*, Melbourne, Oxford University Press

FEDIS, Jaarverslag: De Distributie in 1995, 1996. 1997, Brussels

Fiorino, D. (1995) *Making Environmental Policy*, University of California Press, Berkeley

Fiorino, D. (1996) Toward a new system of environmental regulation: the case for an industry sector approach, *Environmental Law*, 26, pp. 457-488

Fiorino, D. (1996) Environmental policy and the participation gap, pp. 194-212, in Lafferty and Meadowcroft (eds.), op.cit.

Forester, J. (1993) *Critical Theory, Public Policy, and Planning Practice. Towards a critical pragmatism*, State University of New York Press, Albany

Freeman, R and W. Evan (1990) Corporate governance: a stakeholder interpretation, *Journal of Behavioural Economics*

Friedmann, J. (1989) *Planning in the Public Domain; From Knowledge to Action*, New Jersey, Princeton

Fukuyama, F. (1989) Das Ende der Geschichte, *Europäische Rundschau*, vol. 17, no. 4, pp. 3-25

Gamble, A. (1988) *The Free Economy and the Strong State: The Politics of Thatcherism*, Macmillan, London

Gamble, A and G. Kelly (1996) Stakeholder capitalism and one nation socialism, *Renewal* 4, pp. 23-32

Glacken, C. (1967) *Traces on the Rhodian Shore. Nature and Culture in Western Thought from Ancient Times to the End of the Eighteenth Century*, University of California Press, Berkeley

Glasbergen, P. (1992) Seven steps towards an instrumentation theory for environmental policy, *Policy and Politics*, vol. 20, no. 3, pp. 191-200

Glasbergen, P. and P. Driessen (1993) New strategies for environmental policy, *Dutch Crossing* 50, pp. 28-45

Glasbergen, P. (1994, 4th edn.) Milieuproblemen als beleidsvraagstuk, in Glasbergen, P. (ed.) *Milieubeleid. Een beleidswetenschappelijke inleiding*, pp. 15-31. VUGA, 's-Gravenhage

Glasbergen, P. (ed) (1995) *Managing environmental disputes. Network management as an alternative*, Kluwer Academic Publishers, Dordrecht

Glasbergen, P. and R. Cörvers (1995) Environmental problems in an international context, pp. 1-29 in P. Glasbergen and A. Blowers (eds.), op.cit.

Glasbergen, P. and A. Blowers (eds.) (1995) *Environmental Policy in an International Context*, Book 1, *Perspectives*, Arnold

Glasbergen, P. (1996) From regulatory control to network management, in G. Winter (ed), *European environmental law. A comparative perspective*, Aldershot, Dartmouth, pp. 185-200

Glasbergen, P. (1996) Learning to manage the environment, pp. 175-193 in Lafferty and Meadowcroft (eds.), op.cit.

Glasl, F. (1994, 4th edn.) *Konfliktmanagement. Ein Handbuch für Führungskräfte und Berater*, Haupt, Berne

Glasl, F. and E. Brugger (eds.) (1994) *Der Erfolgskurs Schlanker Unternehmen. Impulstexte und Praxisbeispiele*, Manz Verlag, Vienna

Glasl, F. and B. Lievegoed (1993) *Dynamische Unternehmensentwicklung. Wie Pionierbetriebe und Bürokratien zu Schlanken Unternehmen werden*, Haupt, Berne

Goudie, A. (1986) *The Human Impact on the Natural Environment*, Basil Blackwell, Oxford

Goverde, H.J.M. (1993) Verschuivingen in het milieubeleid: van milieuhygiëne naar omgevingsmanagement, in A.J.A. Godfroij, N.J.M. Nelissen (eds.) *Verschuivingen in de besturing van de samenleving*, pp. 49-87, Coutinho, Bussum

Graaf, H. van de, R. van Est en J. Eberg (1996) Beleidsveranderingen beleidsgericht leren, in J. Eberg et al. (red.). *Leren met beleid*, Het Spinhuis, Amsterdam, pp. 1-14

Grabherr, G. (ed.) (1993) *The Embedded Firm. On the socio-economics of industrial networks*, Routledge, London

Gray, B. (1989) *Collaborating. Finding common ground for multiparty problems*, Jossey-Bass Publishers

Green, K., A. McMeekin, A. Irwin (1994) Technological Trajectories and R&D for Environmental Innovation in UK Firms, *Futures* 1994, 26 (10), pp. 1047 - 1059

Grene, M. (1978) Paradoxes of historicity, *Review of Metaphysics*, 32, 15-36

Grene, M. (1985) Perception, interpretation, and the sciences. Toward a new philosophy of science, in D.J. Depew and B.H. Weber (eds.), *Evolution at a Crossroads. The New Biology and the New Philosophy of Science* (pp. 1-20), MIT Press, Cambridge, Ma.

Grene, M. (1987) Hierarchies in biology, *American Scientist*, 75, pp. 504-510

Grossekettler, H. (1991) Zur theoretischen Integration der Finanz- und Wettbewerbspolitik in die Konzeption des ökonomischen Liberalismus, in E. Boettcher et al. (Hrsg.), *Jahrbuch für Neue Politische Ökonomie*, Band 10, Tübingen, pp. 103 - 144

Grubb, M. (1993) *The Earth Summit Agreements: A Guide and Assessment*, Earthscan, London

Gunderson, L., C. S. Holling, and S. Light,(Eds.). (1995) *Barries and Bridges to the Renewal of Ecosystems and Institutions*, Columbia University Press, New York

Gunningham, N. (1995) Environment, self-regulation, and the chemical industry: assessing responsible care, *Law and Policy*, vol. 17, no. 1, pp. 57-109

Gunsteren, H. van (1994) *Culturen van besturen*, Boom

Habermas, J. (1976) *Legitimation Crisis, Heinemann*, London

Habermas, J. (1981) *Theorie des Kommunikativen Handelns I+II*, Suhrkamp Verlag, Frankfurt

Habermas, J. (1987b, 4th edn.) *Theorie des kommunikativen Handelns. Vol. 2: Zur Kritik funktionalistischer Vernunft*, Suhrkamp, Frankfurt am Main

Habermas, J. (1992) *Faktizität und Geltung: Beitrage zur Diskurstheorie des Rechts und des demokratischen Rechtsstaats*, Suhrkamp Verlag, Frankfurt

Hacking, I. (1997) *Taking bad arguments seriously*, London Review of Books, 19 (16), pp. 14-16

Haila, Y., T. Ryynänen and M. Saraste (1971) *Ei vettä rantaa rakkaampaa. Puunjalostusteollisuus vesiemme pilaajana*, Weilin+Göös, Helsinki

Haila, Y., and R. Levins (1992) Humanity and Nature. *Ecology, Science and Society*, Pluto Press, London

Haila, Y. (1992) Measuring nature: quantitative data in field biology, in A.E. Clarke and J.H. Fujimura (eds.), *The Right Tools for the Job. At Work in Twentieth-Century Life Sciences*, pp. 233-253, Princeton University Press, Princeton

Haila, Y. (1994) Preserving ecological diversity in boreal forests: ecological background, research, and management, *Annales Zoologici Fennici*, 31, pp. 203-217

Haila, Y., and L. Heininen (1995) Ecology. A new discipline for disciplining? *Social Text*, 42 (Spring 1995), pp. 153-171

Haila, Y. (1995) Natural dynamics as a model for management: is the analogue practicable?, in A.L. Sippola et al. (eds.), *Northern Wilderness Areas: Ecology, Sustainability*, Values (pp. 9-26), Rovaniemi, Arctic Centre Publications 7

Haila, Y., and C.R. Margules (1996) Survey research in conservation biology, *Ecography*, 19, pp. 323-331

Haila, Y. (1997) Discipline or solidarity? Ecology as politics, in P. J. Taylor, P. Edwards, and S. Haftorn (eds.), *Changing Life: Genomes - Ecologies - Bodies - Commodities*, Minneapolis-St. Paul, University of Minnesota Press

Hajer, M.A. (1995) *The Politics of Environmental Discourse: Ecological Modernisation and the Policy Process,* Oxford University Press, Oxford

Hajer, M.A. (1996) Ecological modernisation as cultural politics, in S. Lash et al., *Environment and Modernity: Towards a New Ecology*, Sage, pp. 246 -68, London

Hanf, K and F.W. Scharpf (1975) *Modernisierung der Volkswirtschaft. Technologiepolitik als Strukturpolitik*, Frankfurt am Main

Hannigan, J.A. (1995) *Environmental Sociology: A Social Constructionist Perspective*, Routledge, London

Haraway, D. (1991) *Simians, Cyborgs, and Women. The Reinvention of Nature*, Routledge, London

Harmon, M.M. (1981) *Action Theory for Public Administration*, Longman, New York

Haury, S. (1989) *Laterale Kooperation zwischen Unternehmen: Erfolgskriterien und Klippen*, Rüegger, Grüsch

Hawkins, K. (1984) *Environment and Enforcement: Regulation and the Social Definition of Pollution*, Clarendon Press, Oxford

Hayles, N.K. (1995) Searching for common ground, in M.E. Soulé and G. Lease (eds.), *Reinventing Nature. Responses to Postmodern Deconstruction* (pp. 47-63), Island Press, Washington D.C.

Hayward, B. (1995) The greening of participatory democracy. a reconsideration of theory, *Environmental Politics*, vol.4, no.4, pp.215-236 (Special Issue: Ecology and Democracy)

Helm, D. (1991) (ed.) *Economic Policy Towards the Environment*, Oxford University Press, Oxford

Hemmelskamp, J. et al. (1995) *The Impact of Parameters provided by Environmental Policy on the Innovative Behaviour of Companies in selected European Countries. Final Report of a Study commissioned by the European Commission DG XII*, Contract Nr. EV5V-CT94-o376, Mannheim, Amsterdam

Henneke, H.G. (1991) *Informelles Verwaltungshandeln im Wirtschaftsverwaltungsrecht und Umweltrecht*, NuR, pp. 267 - 275

Hennis, W., P. Kielmannsegg, and U. Matz (eds.) (1979) *Regierbarkeit. Studien zu ihrer Problematisierung*, vol.2, Stuttgart

Hirst, P. (1994) *Associative Democracy*, Polity Press, Cambridge

HMSO (1995) *Review of Radioactive Waste Management Policy*, London, CM 2919

HMSO (1992) *The UK Environment*, London

HMSO (1990) *This Common Inheritance*, London, CM 1200

Hoekema, A.J. (1994) Enkele inleidende vragen over onderhandelend bestuur, in H.D Stout en A.J. Hoekema (red), *Onderhandelend bestuur*, W.E.J. Tjeenk Willink, Zwolle

Hoffmann-Riem, W. (1990) *Reform des allgemeinen Verwaltungsrechts als Aufgabe - Ansätze am Beispiel des Umweltschutzes*, AöR, pp. 400 - 447

Hoffmann-Riem, W. and E. Schmidt-Aßmann (eds.) (1990) *Konfliktbewältigung durch Verhandlungen*, vol.1 Informelle und mittlerunterstützte Verhandlungen in Verwaltungsverfahren. vol.2 Konfliktmittelung in Verwaltungsverfahren. Nomos, Baden-Baden

Hohmeyer, O., H. Koschel (1995) *Umweltpolitische Instrumente zur Förderung des Einsatzes integrierter Umwelttechnik. Gutachten im Auftrag des Büros für Technikfolgenabschätzung beim Deutschen Bundestag*, Mannheim

Holling, C.S. (1973) Resilience and stability of ecological systems, *Annual Review of Ecology and Systematics*, 4, pp. 1-24

Holling, C.S. (1986) Resilience of ecosystems, local surprise and global change, in W.C. Clark and R. E. Munn (eds.), *Sustainable Development of the Biosphere*, pp. 292-317, Cambridge University Press, Cambridge

278 *Bibliography*

Holling, C.S., and G.K. Meffe (1996) Command and control and the pathology of natural resource management, *Conservation Biology*, 10, pp. 328-337

Holm, J. and B. Klemmensen (1993) *Brancheaftaler - baggrund og betydning*, paper presented at Nordisk Statsvetarkongres, Oslo, 19-21 August 1996

Holzhey, M., H. Tegner (1996) Selbstverpflichtungen - ein Ausweg aus der umweltpolitischen Sackgasse?, *Wirtschaftsdienst*, 1996/VIII, pp. 425 - 430

Homburger, B. (1996) Altauto-Recycling endlich auf dem Weg. Erklärung der umweltpolitischen Sprecherin der F.D.P.-Bundestagsfraktion vom 21.2.1996

Hoorick G. van and C. Lambert (1995) Het decreet betreffende de Milieubeleidsovereenkomsten, *Tijdschrift voor Milieurecht*, pp. 2-10

Hukkinen, J. (1995) Corporatism as an impediment to ecological sustenance: The case of Finnish waste management, *Ecological Economics*, 15, pp. 59-75

Huls, N.J.H. en H.D. Stout (red.) (1992) *Reflexies op reflexief recht*, W.E.J. Tjeenk Willink

HWV Olten (ed.) (1995) *Conference documentation: New Public Management*, June 1995, Höhere Wirtschafts- und Verwaltungsschule (HWV) Olten

IEA (1993) *Electricity Info 1993*, Paris

IEA (1993) *Energy Policies of IEA Countries*, Paris

Ingold, T. (1992) Culture and the perception of the environment, in E. Croll and D. Parkin (eds.), *Bush Base: Forest Farm. Culture, Environment and Development* (pp. 39-56), Routledge, London

Ingold, T. (1993) Globes and spheres. The topology of environmentalism, in K. Milton (ed.), *Environmentalism. The View from Anthropology*, pp. 31-42, Routledge, London

Ingold, T. (ed.). (1996) Key Debates in Anthropology, Routledge, London

IPCC - Intergovernmental Panel on Climate Change (1995a) *Second Assessment Synthesis of Scientific-technical Information Relevant to Interpreting Article 2 of the UN Framework Convention on Climate Change*, Genf.

Jacobs, M. (1995) Sustainability and 'the market': A typology of environmental economics, pp. 46-70 in R. Eckersley (ed.), op.cit.

Jänicke, M. (1992) Conditions for Environmental Policy Success: an International Comparison, *The Environmentalist* 12, 1, 47-58

Jänicke, M., H. Monch and M. Binder (1993) Ecological aspects of structural change, *Intereconomics* 28, pp. 159-169

Jänicke, M. and H. Weidner (eds.) (1995) *Successful Environmental Policy. A critical evaluation of 24 cases*, Sigma

Jänicke, M. And H. Weidner (1997) Summary: global environmental policy learning, in Jänicke and Weidner (eds.), Springer, pp. 299-309, op.cit.

Jänicke, M. and H. Weidner (1997) (eds.), *National Environmental Policies. A comparative study of capacity building*, Springer, Berlin

Jokinen, P. (1997) Agricultural policy community and the challenge of greening: The case of Finnish agri-environmental policy, *Environmental Politics*, 6, pp. 48-71

Kamiske, G.F. and T. Füermann (1995) Reengineering versus Prozessmanagement, in *IndexCompact*, pp. 8-9, Heft 3

Kellert, S. H. (1993) *In the Wake of Chaos. Unpredictable Order in Dynamical Systems*, University of Chicago Press, Chicago

Kettner, M. (1994) Geltungsansprüche, in G. Meggle, and U. Wessels (eds.) *Analyomen 1. Proceedings of the 1st Conference Perspectives in Analytical Philosoph*, de Gruyter, Berlin

Kettner, M. (1995) Habermas über die Einheit der praktischen Vernunft. Eine Kritik, in A. Wüstehube (ed.) *Rationalität*, Königshausen, Neumann, Würzburg

Kickert, W.J.M, E-H. Klein and J.F.M. Koppejan (1997) Managing networks in the public sector: findings and reflections, in W.J.M. Kickert, E-H. Klein and J.F.M. Koppenjan, *Managing complex networks. Strategies for the Public Sector*, SAGE Publications, pp. 166-1991

King, D. (1987) *The New Right: Policy, Markets and Citizenship*, Macmillan, London

Kirkhart, L. (1971) Towards a theory of public administration, in F. Marini (ed.) *Towards a New Public Administration*, pp. 127-164, Chandler, San Francisco

Kjellerup, U. (1995) "Kejserens nye klæder" - Aftaler i Miljøretten, pp. 207-230 in Lin Adrian et al., Ret & privatisering, GadJura, København

Klok, P-J. (1989) *Convenanten als instrument van milieubeleid*, Universiteit Twente, Enschede

Kluegel, J.R., D.S. Mason and B. Wegener (1995) *Social Justice and Political Change: Public Opinion in Capitalist and Post-Communist States*, Aldine de Gruyter, New York

Knoepfel, P. (1994) Erschliessung neuer Instrumente für die Umweltpolitik von morgen durch Intra- und Interpolicy-Kooperation, in *Cahier de l'IDHEAP*, no. 112a. EPF, Lausanne

Knoepfel, P. And I. Kissling-Näf (1993) Transformation öffentlicher Politiken durch Verräumlichung Betrachtungen zum gewandelten Verhältnis zwischen Raum und Politik, *Politische Vierteljahresschrift*, vol. 34, Sonderheft 24: Policy-Analyse. Kritik und Neuorientierung, pp. 267-288

Kohlhaas, M. and B. Praetorius (1994) *Selbstverpflichtungen der Industrie zur CO_2-Reduktion*, Berlin

Konijnenbelt, W. (1992) *Convenanten met de gemeente: fluiten in het schemerduister*, Uitgeverij Lemma

Kooiman, J. (1993) (ed.), *Modern Governance, new government-society interactions*, Sage, London

Kooiman, J. (1993) Governance and governability: using complexity, dynamics and diversity, pp. 35-48, in Kooiman (ed.), op.cit.

Koschel, H. And S. Weinreich (1995) Ökologische Steuerreform auf dem Prüfstand - Ist die Zeit reif zum Handeln?, in O. Hohmeyer (Hrsg.), *Ökologische Steuerreform*, ZEW Wirtschaftsanalysen Bd. 1, pp. 9 - 38

Krause, F., J. Koomey, D. Olivier (1994) Incorporating global warming ex-ternalities through environmental least cost plannning: a case study of Western Europe, in O. Hohmeyer, R. L. Ottinger, *Social Costs of Energy*, Berlin, Heidelberg, New York, pp. 287 - 312

Kreuzberg, P. (1993) Zur ökonomischen Rationalität freiwilliger Kooperationslösungen für das Klimaproblem, *Zeitschrift für Energiewirtschaft* (ZfE) 4/93, pp. 304 - 309

Lafferty, W.M. and J. Meadowcroft (1996) Democracy and the environment: congruence and conflict - preliminary reflections, pp. 1-17 in Lafferty and Meadowcroft (eds.), op.cit.

Lafferty, W.M. and J. Meadowcroft (eds.) (1996) *Democracy and the Environment. Problems and prospects*, Edward Elgar

Latour, B. (1993) *We Have Never Been Modern*, Harvard University Press, Cambridge, Ma.

Lauber, V. and K. Hofer (1997) *Business and Government Motives for Negotiating Voluntary Agreements: A comparison of the experiences in Austria, Denmark and the Netherlands*, paper presented at the 25th ECPR Joint Sessions of Workshops, Workshop: 'The effectiveness of policy instruments for improving EU environmental policy', Bern, 27 February-4 March 1997

Le Grand, J. (1993) Ein Wandel in der Verwendung von Policy-Instrumenten. Quasi-Märkte und Gesundheitspolitik, in A. Héritier (ed.) *Policy-Analyse. Kritik und Neuorientierung*, pp. 225-244. Special issue 24/1993 Politische Vierteljahresschrift (PVS), vol. 34, Westdeutscher, Opladen

Lee, K. N. (1993) *Compass and Gyroscope. Integrating Science and Politics for the Environment*, Island Press, Washington D.C.

Lehtinen, A.A. (1991) *Northern natures. A study of the froest question emerging within the timer-line conflict in Finland*, Fennia, 169, pp. 57-169

Lehtinen, A.A. (1995) The forest question - a challenge for the state and a concern for civil society, in L. Granberg and J. Nikula (eds.), *The Peasant State. The State and Rural Question in 20th Century Finland*, pp. 121-132, Rovaniemi, University of Lapland Publications in the Social Sciences

Leroy, P. (1994) De ontwikkeling van het milieubeleid en de milieubeleidstheorie, pp. 35-58 in P. Glasbergen (ed.) op.cit.

Levins, R. (1996) Ten propositions on science and antiscience, in A. Ross (ed.), *Science Wars*, Duke University Press, Durham, pp. 180-191

Lewis, S. (1996) *Voluntary agreements in the US*, paper presented at the conference 'The Economy and Law of Volunatry Approaches in Environmental Policy' in Fondazione Levi, Venezia, 18-19 November 1996

Lewis, M. W. (1992) *Green Delusions. An Environmentalist Critique of Radical Environmentalism*, Duke University Press, Durham

Liefferink, D. and A. Mol (1996) *Voluntary agreements as a form of deregulation: The Dutch experience*, paper presented at the conference 'Deregulation and the Environment' in Badia Fiesolana, Florence, 9-11 may 1996

Limacher, P and J. Münch (1995) *Policy Network und New Public Management in der Umweltpolitik*, Working paper on Quantitative Geography and Human ecology. No. 3, Department of Geography ETH, Zurich

Lindblom, C. (1977) *Politics and Markets: The World's Political-Economic Systems*, Basic Books, New York

Lindblom, C (1979) Still muddling, not yet through, *Public Administration Review* 39, pp. 517-526

List, M. (1996) Sovereign States and International Regimes, pp. 7-23, in Blowers and Glasbergen, op.cit.

Löfgren, K.G. (1995) Markets and externalities, in H. Folmer, H. Landis Gabel and H. Opschoor, *Principles of Environmental and Resource Economics. A guide for students and decision-makers*, Edward Elgar, pp. 17-46

Lorenzen, K. H. et al. (1994) Voluntary strategies in pollution prevention: experiences from Denmark and the Netherlands, *European Environment*, vol.4, part 4, august 1994, pp. 18-22

Luhmann, N. (1988) *Die Wirtschaft der Gesellschaft*, Suhrkamp, Frankfurt am Main

Luhmann, N. (1990) *Essays on selfreference*, New York

MacIntyre, A. (1985) *After Virtue. A Study in Moral Theory* (2nd ed.), Duckworth, London

Maier-Rigaud, G. (1995) Für eine ökologische Wirtschaftsordnung, *Jahrbuch Ökologie* 1996 München.

Marin, B. and R. Mayntz. (1991) (eds.), *Policy Networks*, Campus Verlag, Frankfurt

Marsh, G. P. (1965) *Man and Nature. Or, Physical Geography as Modified by Human Action*, (Edited by David Lowenthal; original in 1864), Belknap Press, Cambridge, Ma.

Marsh, D. and R. Rhodes (1992) *Policy Networks in British Government*, Oxford University Press, Oxford

Mathews, F. (1995) Community and the Ecological Self, *Environmental Politics*, vol.4, no.4, pp. 66-100 (Special Issue: Ecology and Democracy)

Mayntz, R. (1978) *Vollzugsprobleme der Umweltpolitik*, Stuttgart

Mayntz, R. (1993) Governing failures and the problem of governability: some comments on a theoretical paradigm, pp. 9-20, in J. Kooiman (ed.) op.cit.

Meadowcroft, J. (1997a) Planning, democracy and the challenge of sustainable development, *International Political Science Review* 18, pp. 167-190

Meadowcroft, J. (1997b) Planning for sustainable development: insights from the literatures of political science, *European Journal of Political Science Annual Review*

Meadows, D.L. (1972) *The Limits to Growth. A Report for the Club of Rome Project on The Predicament of Mankind*, Universe Books, New York

Merchant, C. (1992) *Radical Ecology. The Search for a Livable World*, Routledge, New York

Metzen, H. (1994) *Schlankheitskur für den Staat. Lean Management in der öffentlichen Verwaltung*, Campus, Frankfurt am Main

Mez, L. (1995) Erfahrungen mit der ökologischen Steuerreform in Dänemark, in O. Hohmeyer (Hrsg.), *Ökologische Steuerreform*, ZEW-Wirtschaftsanalysen Band 1, Baden-Baden, pp. 109 - 128

Milton, K. (1996) *Environmentalism and Cultural Theory. Exploring the Role of Anthropology in Environmental Discourse*, Routledge, London

Ministerie van Algemene Zaken (1995) *Aanwijzigingen voor convenanten*, Regeling van 18 december 1995, nr. 95M009543, Den Haag

Ministry of Environment (1994) *Voluntary environmental agreements - scopes, problems and perspectives. Letter from the Danish Environmental Protection Agency to the European Commission*

Mol, A. and G. Spaargaren (1993) Environment, modernity and the risk society: the apocalyptic horizon of environmental reform, *International Sociology* 8, pp. 432-459

Mol, A. (1995) *The Refinement of Production : Ecological Modernization Theory and the Chemical Industry*, van Arkel, Utrecht

Mol, A. (1996) Ecological modernisation and institutional reflexivity: environmental reform in the late modern age, *Environmental Politics*, vol.5, no.2, Summer, pp. 302 - 323

Moran, A. (1995) Tools of environmental policy: market instruments versus command-and-control, in R. Eckersley (ed.), *Markets, the State and the Environment. Towards integration*, MACMILLAN, pp. 73-85

Müggenborg, H.-J. (1990) *Formen des Kooperationsprinzips im Umweltrecht der Bundesrepublik Deutschland*, NVwZ, pp. 909 - 917

Muller, F. (1995) *Energy Taxes, the Climate Change Convention and Economic Competitiveness*, Paper Presented to the 3rd International Conference on Social Costs, Ladenburg, Germany, 27 - 30 May

Murphy, J.B. (1994) The kinds of order in society, in P. Mirowski (ed.), *Natural Images in Economic Thought. Markets Read in Tooth & Claw*, Cambridge University Press, Cambridge, pp. 536-582

Murswiek, D. (1988) Freiheit und Freiwilligkeit im Umweltrecht, *Juristen Zeitung*, 43. Jg., H. 21, pp. 985 - 993

Nelissen, N.J.M. (1990) *Afscheid van de vervuilende samenleving*, Zeist

Nicolis, G., and I. Prigogine (1989) *Exploring Complexity. An Introduction*, New York, W.H. Freeman and Co

Nirex (1997) *Science Report: Nirex Approach to Publication and Peer Review*, Harwell, UK Nirex Ltd., Report No. S/97/001, January

OECD (1991) *The State of the Environment*, Paris.

OECD, GEEPI, *Evaluating the Efficiency and Effectiveness of Economic Instruments in Environmental Policy*, Paris, May 13, 1996

Ökologische Briefe (1996) (5): *Europäische Kommission will das Fünfliterauto*

Ökologische Briefe (1996) (2): *Denkhilfe" für die Wirtschaft aus dem Umweltbundesamt*

Osborne, D. and T. Gaebler (1992) *Reinventing Government*, Reading, MA, Addison/Wesley

Ostrom, E. (1990) *Governing the Commons*, Cambridge University Press, Cambridge *Overeenkomst betreffende Emissiereducties van SO$_2$ en NO$_x$ afkomstig van Elektriciteitsproduktie-installaties*

O'Connor, J. (1973) *The Fiscal Crisis of the State*, St. Martin's Press, New York

O'Neill, R. V., et al. (1986) *A Hierarchical Concept of Ecosystems*, Princeton Univ. Press, Princeton, N.J.

O'Riordan, T., R. Kemp and M. Purdue (1988) *Sizewell 'B': Anatomy of the Inquiry*, Macmillan, London

Paehlke, R. (1989) *Environmentalism and the Future of Progressive Politics*, Yale University Press, New Haven

Paehlke, R. (1996) Environmental challenges to democratic practice, pp. 18-38, in Lafferty and Meadowcroft (eds.), op.cit.

Paehlke, R. (1997) Environmental values and public policy, in N. Vig and M. Kraft (eds.), *Environmental Policy in the 1990s* (third edition), Congressional Quarterly Press, Washington, D.C.

Pakulski, J. (1991) *Social Movements: The Politics of Moral Protest*, Longman Cheshire, Melbourne

Palmer, K., W.E. Oats and P.R. Portney (1995) Tightening environmental standards: the benefit-cost or the no-cost paradigm?, *Journal of Economic Perspectives*, vol. 9, no. 4, pp. 119-132

Pearce, D.W. and R.K. Turner (1990) *Economics of Natural Resources and the Environment*, Harvester Wheatsheaf

Pearce D.W. and Warford (1993) *World Without End: Economics, Environment and Sustainable Development*, Oxford University Press, Oxford

Peppel, R.A. van de (1995) *Naleving van het milieurecht. Toepassing van beleidsinstrumenten op de Nederlandse verfindustrie*, Kluwer, Deventer

Peters, J. (1993) Voluntary agreements between government and industry: The basic metal covenant as an example, pp. 19-32 in van Dunné (ed.), op.cit.

Piore, M. and Ch. Sabel (1984) *The second industrial divide. Possibilities for Prosperity*, Basic Books, New York

Plumwood, V. (1995) Has democracy failed ecology? An eco-feminist perspective, *Environmental Politics*, vol.4, no.4, pp. 134-168 (Special Issue: Ecology and Democracy)

Polsby, N. (1985) Prospects for pluralism, *Society* 22, pp. 30-34

Ponting, C. (1991) *A Green History of the World*, Penguin London

Porter, M.E. and C. van der Linde (1995a) Green and competitive: ending the stalemate, *Harvard Business Review*, september-october, pp. 120-134

Porter, M.E. and C. van der Linde (1995b) Toward a new conception of the environment - competitiveness relationship, *Journal of Economic Perspectives*, vol. 9, no. 4, pp. 97-118

Potter, D. (1996) Non-governmental organisations and environmental policies, in Blowers and Glasbergen (eds.), *Environmental Policy in an International Context: Prospects*, Arnold, London, pp. 25 - 49

Potter, D. and Taylor, A. (1996) Introduction, in D. Potter (ed.), *NGOs and Environmental Policies: Asia and Africa*, Frank Cass, London

Poulsen, J. (1986) *The Holistic Conception of Democracy*, Dissertation, Yale

Princen, T. and M. Finger (1994) *Environmental NGOs in World Politics, Routledge*, London

Pringle, P. and J. Spigelman (1982) *The Nuclear Barons*, London, Sphere Books. Radioactive Waste Management Advisory Committee (RWMAC)(1997) *Response on: Nirex's Proposals for Publication and Peer Review*, London, January

Prittwitz, V. von (ed.) (1993) *Umweltpolitik als Modernisierungsprozeß. Politikwissenschaftliche Umweltforschung und -lehre in der Bundesrepublik*, Leske + Budrich, Opladen

Quine, W. (1953) *From a Logical Point of View*, Harvard University Press, Cambridge, Ma.

Rabe, B. (1988) The politics of environmental dispute resolution, *Policy Studies Journal* 16, pp. 565-601

Radkau, J. (1997) The wordy worship of nature and the tacit feeling for nature in the history of German forestry, in M. Teich, R. Porter, and B. Gustafsson (eds.), *Nature and Society in Historical Context*, Cambridge University Press, Cambridge, pp. 228-239

Rannikko, P. (1995) Restructuring of forestry and forest villages in eastern Finland, in L. Granberg and J. Nikula (eds.), *The Peasant State. The State and Rural Question in 20th Century Finland*, University of Lapland Publications in the Social Sciences, Rovaniemi, pp. 109-118

Raumolin, J. (1984) *Metsäsektorin vaikutus Suomen taloudelliseen ja yhteiskunnalliseen kehitykseen* (English summary: The impact of forest sector on economic and social development in Finland), Research Institute of Northen Finland, C51, University of Oulu

Raumolin, J. (1985) *The impact of forest sector on economic development in Finland and eastern Canada*, Fennia, 163, pp. 387-394

Rawls, J. (1962) Justice as Fairness, in *Philosophy, Politics, and Society, second series*, edited by Laslett and Runciman, Basil Blackwell, Oxford, pp. 132-157

Rawls, J. (1988) *A Theory of Justice*, Oxford University Press (first published 1972), Oxford

Reade, E. (1997) Planning in the future or planning of the future?, in A. Blowers, and B. Evans, *Town Planning into the 21st. Century*, Routledge, London, pp. 71 -103

Rees, J.V. (1994) *Hostages of each other. The transformation of nuclear safety since Three Mile Island*, University of Chicago Press

Renn, O., Th. Webler and H. Rakel (1993) Public participation in decision making: A three-step procedure, *Policy Sciences*, vol. 26, pp. 189-214

Renn, O. (1994a) Sozialverträglichkeit der Technikentwicklung, *Österreichische Zeitschrift für Soziologie*, vol. 19, no. 4, pp. 34-49

Renn, O. (1994b) *A Regional Concept of Qualitative Growth and Sustainability Pilot project for the German state of Baden-Württemberg*, Academy of Technology Assessment in Baden-Württemberg, Report no. 2

Rennings, K. (1994) *Indikatoren für eine dauerhaft-umweltgerechte Entwicklung*, Stuttgart

Rennings, K., K.L. Brockmann, H. Koschel, H. Bergmann, I. Kühn (1996) *Nachhaltigkeit, Ordnungspolitik und freiwillige Selbstverpflichtung*, Heidelberg

Rhodes, R. (1994) The hollowing out of the state: the changing nature of the public service in Britain, *The Political Quarterly*

Rhodes, R. (1996) The new governance: governing without government, *Political Studies* 44, pp. 652-667

Rijkinstituut voor Volksgezondheid en Milieuhygenie (RIVM) (1989) *Concern for Tomorrow*, Bilthoven

Ritsema van Eck, E.J. en P. Huguenin (1993) *Conflicthantering en onderhandelen; een praktische inleiding*, Bohn Stafleu Van Loghum, Houten/Zaventem

Rittel, H. and M. Weber (1973), Dilemmas in a general theory of planning, *Policy Sciences* 4, pp. 155-169

Rokkan, S. (1966) Numerical democracy and corporate pluralism, in R. Dahl (ed.), *Political Oppositions in Western Democracies*, Yale University Press, New Haven

Rosen, R. (1977) Observation and biological systems, *Bulletin of Mathematical Biology*, 39, pp. 663-678

Rosen, R. (1989) Similitude, similarity, and scaling, *Landscape Ecology*, 3, pp. 207-216

Royal Society, The (1992) *Risk: Analysis, Perception, Management*, London

Sabatier, P.A. and H.C. Jenkins-Smyth (1993) *Policy Change and Learning: An Advocacy Coalition Approach*, Westview Press, Boulder, Colorado

Sacksofsky, E. (1996) Anmerkungen zu verschiedenen Konzepten einer Neuregelung der Altautoentsorgung, *Zeitschrift für Umweltpolitik & Umweltrecht*, Jg. 19, Heft 1, S. 99-108

Sage, C. (1996) Population, poverty and land in the South, in P. Sloep and A. Blowers, *Environmental Policy in an International Context: Conflicts*, Arnold, London, pp. 97 - 126

Salthe, S.N. (1985) *Evolving Hierarchical Systems*, Columbia University Press, New York

Sanghi, A.K. (1995) *Climate for Climate Change Actions in the U.S.: The New York Experience*, Paper Presented to the 3rd International Conference on Social Costs, Ladenburg, Germany, 27 - 30 May

Schaltegger, et al. (1996) *Innovatives Management staatlicher Umweltpolitik. Das Konzept des New Public Environmental Management*, Birkhäuser, Basel

Scharpf, F.W. (1991) Die Handlungsfähigkeit des Staates am Ende des zwanzigsten Jahrhunderts, *Politische Vierteljahresschrift*, vol. 32, pp. 621-634

Schedler, K. (1995) *Ansätze einer wirkungsorientierten Verwaltungsführung. Von der Idee des New Public Managements (NPM) zum konkreten Gestaltungsmodell*, Fallbeispiel Schweiz, Haupt, Berne

Schmidt, A. (1995) Der möglichen Beitrag der Kooperation zum Innovationserfolg fürt kleine und mittelgrosse Unternehmen (KMU), *Zeitschrift für Betriebswirtschaft*, Additional issue 1, pp. 121-149

Schmitter, P. and K. Lembruch (1979) (eds.), *Trends Toward Corporatist Intermediation*, Sage, New York

Schnaiberg, A. (1980) *The Environment: From Surplus to Scarcity*, Oxford University Press

Schuppert, G.F. (1989) Zur Neubelebung der Staatsdiskussion. Entzauberung des Staates oder bringing the state back in?, *Der Staat*, vol. 28, nr. 1, pp. 93-104.

Self, P. (1985) *Political Theories of Modern Government. Its role and reform*, George Allen and Unwin

Sen, A. (1993) Capability and well-being, in M. C. Nussbaum and A. Sen (eds.), *The Quality of Life*, Clarendon Press, Oxford, pp. 30-53

Sengenberger, W. and F. Pyke (1992) Industrial districts and local economic regeneration: Research and policy issues, in F. Pyke, and W. Sengenberger, (eds.) Industrial Districts and Local Economic Regeneration, pp. 3-29, *International Institute for Labour Studies*, Geneva

Seyad A. (1994) *Convenanten als instrument van milieubeleid in de verpakkingssector*, Gent

Seyad A., F. Senesael, M. De Clercq (1995) *Voluntary Collection & Recycling Initiatives for Packaging Waste: Critical Evaluation of the Belgian FOST Plus System*, Gent

Seyad, A., M. de Clerq, F. Senesael (1996) *The Use of Voluntary Agreements as Instruments of Environmental Policy in the Energy Sector - the Belgian Electricity Voluntary Agreement Case*, Paper presented on the Sixth Annual Greening of Industry Network Conference, Heidelberg, November 1996

Seyad A., S. Baeke, M. De Clercq (1998) Vrijwillige inzameling en recyclage van verpakkingsafval in België, in K.R.D. Lulofs and G.J.I. Schrama (red.), Ketenbeheer, CSTM Jaarboek 1998, Twente University Press, Enschede

Silverman, D. (1971) *The Theory of Organisations*, Basic Books, New York Simon, W. and M. Heß, (1989) *Handbuch Qualitäts-Zirkel. Hilfsmittel zur Produktion von Qualität*, Verlag TÜV Rheinland, Köln

Sklair, L. (1994) Global sociology and global environmental change, in M. Redclift, and T. Bento (eds.), *Social Theory and the Global Environment*, Routledge, London, pp. 205 - 27

Slovic, P.B. Fischoff and S. Lichtenstein (1980) Facts and fears: understanding perceived risk, in R. Schwing and W. Albers (eds.), *Societal Risk Assessment: How Safe is Safe Enough?*, Plenum Press , New York

Smith, M. (1993) *Pressure, Power and Policy*, Hemel Hempstead, Harvester Wheatsheaf

Smith, N. (1984) *Uneven Development. Nature, Capital and the Production of Space*, Basil Blackwell, Oxford

Spies, R. (1994) Der 'Grüne Punkt' als ökonomisches Instrument in der Abfallwirtschaftspolitik, in *Zeitschrift für angewandte Umweltforschung*, Jg. 7, Heft 3, pp. 309-321

SRU - Rat von Sachverständigen für Umweltfragen (1994) *Umweltgutachten 1994 - Für eine dauerhaft-umweltgerechte Entwicklung*, Wiesbaden

SRU - Rat von Sachverständigen für Umweltfragen (1996) *Umweltgutachten 1996 - Zur Umsetzung einer dauerhaft-umweltgerechten Entwicklung*, Wiesbaden

Staber, U.H., N.V. Schaefer and B. Sharma (1996) *Business Networks. Prospects for Regional Development*, de Gruyter, Berlin

Stewart-Oaten, A., W.W. Murdoch and K.R. Parker (1986) Environmental impact assessment: 'pseudoreplication' in time? *Ecology*, 67, pp. 929-940

Storsved, A.-S. (1993) The debate on establishing the ministry of the environment in Finland in the light of environmental ideologies, *Environmental Politics*, 2, pp. 304-326

Straaten, J. van der and M. Gordon (1995) Environmental problems from an economic perspective, pp. 133-161, in Glasbergen and Blowers (eds.), op.cit.

Stretton, H. (1976) *Capitalism, Socialism and the Environment*, Cambridge University Press

Susskind, L. and J. Cruikshank (1987) *Breaking the Impasse; Consensual Approaches to Resolving Public Disputes*, Basic Books

Suurland, Jan (1994) The case of Dutch covenants, *European Environment*, vol.4, part 4, august 1994, pp. 3-7

SZ - Süddeutsche Zeitung vom 6.3.1996: Fünf-Liter-Auto gemeinsames Ziel

Szejnwald Brown, H., P. Derr and O. Renn (et al.) (1993) *Corporate Environmentalism in a Global Economy. Societal values in international technology transfer*, London

Talos, E. (1996) Corporatism - the Austrian model, pp. 103-124 in V. Lauber (ed.), *Contemporary Austrian Politics*, Boulder, Westview Press

Taylor, P.J. and F.H. Buttel (1992) How do we know we have global environmental problems? Science and the globalization of environmental discourse, *Geoforum*, 23, pp. 405-416

Taylor, P. J. (1997) Appearances notwithstanding, we are all doing something like political ecology, *Social Epistemology*, 11, pp. 111-127

Taylor, P. and R. García-Barrios (1997) Dynamics and rhetorics of socio-environmental change: Critical perspectives on the limits of Neo-Malthusian environmentalism, *Advances in Human Ecology*, 6, pp. 257-292

Teubner, G.C.M. (1992) Reflexief recht: de kracht van niet-statelijk recht, pp. 71-81, in Huls en Stout (red.), op.cit.

Teubner, G.C.M. (1993) *Law as an autopoietic system*, Oxford

Thielemann, U. (1996) *Das Prinzip Markt. Kritik der ökonomischen Tauschlogik*, Haupt, Bern

Thomas, W. L. J. (ed.) (1956) *Man's Role in Changing the Face of the Earth*, University of Chicago Press, Chicago

Thomas, A. (1996) NGO advocacy, democracy and policy development: some examples relating to environmental policies in Zimbabwe and Botswana, in D. Potter (ed.) *NGOs and Environmental Policies: Asia and Africa*, Frank Cass, London, pp. 38 - 65

Thompson, G., J. Frances, R. Levacic and J. Mitchell (eds.) (1991) *Markets, Hierarchies and Networks. The coordination of social life*, Sage, London

UBA - Umweltbundesamt (1993) *Jahresbericht 1993*, Berlin

Umwelt (1996) Novelle der Verpackungsverordnung, *Umwelt* nr. 4, S. 158

Unipede, 200.03 ENVPOL, Environmental Agreements and Self Commitments, July 1996

United Nations Conference on Environment and Development (UNCED) (1992) *Agenda 21*, New York, United Nations Organisation

VDA - Verband der Automobilindustrie e.V. (1995) *Freiwillige Zusage zur Kraftstoffverbrauchsminderung*, Frankfurt/Main, 22.3.1995

VDEW - Verband Deutscher Elektrizitätswerke (1996) Bericht 1996 zur Erklärung derVDEW zum Klimaschutz, in *BDI - Bundesverband der Deutschen Industrie e.V.: Aktualisierte Erklärung der deutschen Wirtschaft zur Klimavorsorge*, Köln

Vermeulen, W.J.V., J.F.M. van der Waals, P. Glasbergen en H. Ernste (1997) *Duurzaamheid als uitdaging*, Wetenschappelijke Raad voor het Regeringsbeleid, Voorstudie en achtergronden, nr. V101, Den Haag

Vig, N. and M. Kraft (1997) (eds.), *Environmental Policy in the 1990s* (third edition), Congressional Quarterly Press, Washington, D.C.

Virtanen, S. (1995) Terms of nature and modern society in Finland, in M. Seppälä and J. Hiedanpää (eds.), *Layers of Nature and Culture - Luonnon ja kulttuurin kerrostumia*, Pori Art Museum, Pori, pp. 69-82)

Vliet, L.M. van (1992) *Communicative besturing van het milieuhandelen van ondernemingen*, Eburon, Delft

Vliet, M. van (1993) Environmental regulation of business: options and constraints for communicative governance, pp. 105-118, in Kooiman (ed.), op.cit.

Voisey, H., C. Beuermann, L. Sverdrup and T. O'Riordan (1996) The political significance of local Agenda 21: the early stages of some European experience, *Local Environment*, 1, pp. 333-350

Walters, C. J. (1986) *Adaptive Management of Renewable Resources*, McGraw Hill, New York

Weale, A. (1992) *The new politics of pollution*, Manchester University Press, Manchester

Webb, K. and A. Morrison (1996) *Voluntary approaches, the environment, and the law: a Canadian perspective*, paper presented at the conference 'The Economy and Law of Volunatry Approaches in Environmental Policy' in Fondazione Levi, Venezia, 18-19 November 1996

Weber, M. (1972, 5th edn.) *Wirtschaft und Gesellschaft*, Mohr, Tübingen

Webler, T. (1995) 'Right' Discourse in Citizen Participation: An Evaluative Yardstick, in O. Renn, Th. Webler and P. Wiedemann (ed.), *Fairness and Competence in Citizen Participation*, Kluwer Academic Publishers, Dordrecht, pp. 35-86

Weiland, R. (1995) *Rücknahme- und Entsorgungspflichten in der Abfallwirtschaft - Eine institutionenökonomische Analyse der Automobilbranche*, Wiesbaden

Weimer, D. (1995), (ed.) *Institutional Design*, Kluwer Academic, Boston

Welt vom 15.4.1996 Merkel liebäugelt mit Kartellen

Wicke, L. (1989, 2nd edn.) *Umweltökonomie, eine praxisorientierte Einführung*, Vahlen, München

Wiens, J.A. (1996) Oil, seabirds, and science. The effects of the Exxon Valdez oil spill, *BioScience*, 46, pp. 587-597

Willems, W. (1990) Milieuconvenant: opgeblazen instrument?, *Namens*, 2, p. 28

Williams, O.E. (1975) *Markets and Hierarchies. Analysis and Antitrust Implications. A study in the economics of internal organisation*, Free Press, New York

Willke, H. (1983) *Entzauberung des Staates. Überlegungen zu einer sozietalen Steuerungstheorie*, Athenäum, Königstein

Willke, H. (1991) *Systemtheorie*, Stuttgart/New York, 3. Aufl.

Wimsatt, W.C. (1974) Complexity and organization, in K.F. Schaffner and S. Cohen (eds.), *Proceedings of the Meeting of Philosophy of Science Association*, Dordrecht, The Netherlands: Reidel, pp. 67-86

Wimsatt, W.C. (1994) The ontology of complex systems: levels of organization, perspectives, and causal thickets, *Canadian Journal of Philosophy*, 20, pp. 207-274

Winsemius, P. (1989, 5th edn.) *Gast in eigen huis. Beschouwingen over milieumanagement*, Samson Tjeenk Willink, Alphen aan den Rijn

Winsemius P. (1992) *Environmental Contracts and Covenants: New Instruments for a Realistic Environmental Policy?*, pp. 5-15 in Van Dunné (ed.), op.cit.

Witherspoon, S. (1996) Democracy, the environment and public opinion in Western Europe, pp. 39-70, in Lafferty and Meadowcroft (eds.), op.cit.

Woerkum, C. van (1997) *Communicatie en interactieve beleidsvorming*, Bohn Staleu van Loghum

Wolf, E.R. (1982) *Europe and the People without History*, University of California Press, Berkeley

Womack, J.P., D.T. Jones and D. Roos (1992, 7th edn.) *Die zweite Revolution in der Autoindiustrie. Konsequenzen aus der weltweiten Studie des Massachusetts Institute of Technology*, Campus, Frankfurt am Main

Wupperman, B. (1993) Zwei Jahre Verpackungsverordnung: Weitere Machtkonzentration statt Müllvermeidung, *Zeitschrift für angewandte Umweltforschung*, Jg. 6, 4, pp. 448-455

Wynne, B. (1994) Scientific knowledge and the global environment, in M. Redclift and T. Benton (Eds.), *Social Theory and the Global Environment*, Routledge, London, pp. 169-189

Yearley, S. (1989) Bog standards: science and conservation at a public inquiry, *Social Studies of Science*, 19, pp. 421-438

Yearley, S. (1992) *The Green Case: A Sociology of Environmental Issues, Arguments and Politics*, Routledge, London

Yearley, S. (1996) Social movements and environmental change, in M. Redclift and T. Benton (eds.) *Social Theory and the Global Environment*, Routledge, London, pp. 150 -68

Ylikangas, H. (1993) The government and the nation, in M. Rahikainen (ed.), *Austerity and Prosperity. Perspectives on Finnish Society*, University of Helsinki, Lahti Research and Training Centre, Lahti

Zilleßen, H., P.C. Dienel and W. Strubelt (eds.) (1993) *Die Modernisierung der Demokartie. Internationale Ansätze - Westdeutscher Verlag*, Opladen

Zimmermeyer, G. (1995) *Voluntary Fuel Economy Targets for Motor Vehicles in Germany*, Presentation for the IEA/OECD International Workshop on Voluntary Approaches for Mitigating CO_2-Emissions, Bonn

ABOUT THE AUTHORS

PIETER GLASBERGEN is Professor of Environmental Studies; Policy and Management at Utrecht University and the Dutch Open University. He specializes in planning and policy issues, particularly with reference to environmental policy, physical planning, water management, and policy for landscape and nature conservation. Address: Department of Environmental Studies, PO Box 80.115, 3508 TC Utrecht The Netherlands.

STEVEN BAEKE is Licentiate in Applied Economic Sciences. Since 1996 he works as a Researcher at the Centre for Environmental Economics and Environmental Management of the University of Gent in Belgium. He is involved with research on economic instruments of environmental policy, environmental agreements, the packaging problem, and environmental accounting.

ELLEN MARGRETHE BASSE is Professor in Environmental Law and director of the Centre for Social Science Research on the Environment (CeSaM), at the University of Aarhus, Denmark. Her main field of research is legal aspects of governance, especially related to the environmental problematic.

DR. HEIDI BERGMANN is a jurist and holds a diploma in administrative science. Since 1995 she works at the Centre for European Economic Research (ZEW) in Mannheim, Germany. In her scientific work she focusses on environmental and economic law. She conducted several studies dealing with problems of environmental instruments.

ANDREW BLOWERS is Professor of Social Sciences (Planning) at the Open University, UK. Most of his teaching and research has been in the fields of environmental planning, politics and policy. He is particularly concerned with the politics of sustainable development and the problems of radioactive waste.

KARL LUDWIG BROCKMANN is an economist. Since 1994 he works at the Centre for European Economic Research (ZEW) in Mannheim, Germany. The major focus of his work is on the international dimension of environmental problems and on the evolution of environmental policy instruments.

MARC DE CLERCQ is Professor at the University of Gent and Chairman of the Department of General Economics. He is also Chairman of the Centre for Environmental Economics and Environmental Management. Since 1993, he is Chairman of the Follow-Up Commission Ecotaxes (Services of the Belgian Prime Minister) and a member of the group of Experts concerning the Integration of Economic Policy and the Environment.

DR. PETER DRIESSEN is Associate Professor of Environmental Studies at Utrecht University, the Netherlands. Most of his research is related to planning and management issues. He is in particular concerned with the relationships between environmental planning and physical planning and with network management.

MARTIN ENEVOLDSEN is a Ph.D student at the Institute of Political Science, Aarhus University, Denmark, and he is working as a researcher at the Center for Social Research on the Environment (CeSaM). His main research topic is new environmental policy instruments.

DR. HUIB ERNSTE is a Senior Lecturer and Researcher, Department of Geography of the Swiss Federal Institute of Technology, Zurich, Switzerland. From Oct. 1 1998 Head of the Department of Geography, University of Nijmegen, The Netherlands. His main research interests are: action theory, forms of rationality in planning and policy making, cooperative -/communicative planning and management systems for regional sustainable development.

BETTY GEBERS is a lawyer. She co-ordinates the environmental law division at Öko-Institut e.v. (Freiburg/Darmstadt/Berlin), Germany. Her field of work is environmental law, instruments of environmental policy, and environmental management.

YRYÖ HAILA is Professor of Environmental Policy at the University of Tampere, Finland. He has his background in ecology, with main research interest in ecological biogeography and conservation ecology, but has simultaniously worked on philosophical and social dimensions of ecology. Current research circles around the questions, How to avoid the nature-culture dualism when thinking about and acting upon society and the environment?; and What use is ecology?

DR. JAMES MEADOWCROFT is a Senior Lecturer in the Department of Politics at the University of Sheffield (UK). Trained as a political theorist and historian of political thought, his recent work has been focussed on the domain of environmental politics.

DR. KLAUS RENNINGS is an economist. Since 1994 he works at the Centre for European Economic Research (ZEW) in Mannheim, Germany. His main fields of research are the valuation of environmental risks and the operationalization of the concept of sustainable development by monetary and physical indicators.

AKIM SEYAD is Licentiate in Applied Economic Sciences. Since 1994 he works as a Researcher at the Centre for Environmental Economics and Environmental Management of the University of Gent in Belgium. He is involved with research on environmental agreements, economic instruments of environmental policy and the packaging problem.

ENVIRONMENT & POLICY

KLUWER ACADEMIC PUBLISHERS – DORDRECHT / BOSTON / LONDON